S0-EEW-543

IMPACT OF ACID RAIN AND DEPOSITION ON AQUATIC BIOLOGICAL SYSTEMS

A symposium
sponsored by
ASTM Committee D-19
on Water
Bal Harbour, FL, 29 Oct. 1984

ASTM SPECIAL TECHNICAL PUBLICATION 928
Billy G. Isom, Tennessee Valley Authority,
and Sally D. Dennis and John M. Bates,
Ecological Consultants, Inc., editors

ASTM Publication Code Number (PCN)
04-928000-16

 1916 Race Street, Philadelphia, PA 19103

Library of Congress Cataloging-in-Publication Data

Impact of acid rain and deposition on aquatic biological systems.

(ASTM special technical publication; 928)
Includes bibliographies and index.
"ASTM publication code number (PCN) 04-928000-16."
  1. Acid rain—Environmental aspects—Congresses. I. Isom, B. G. (Billy G.) II. Dennis, Sally D. III. Bates, John M. (John Morton), 1932– . IV. American Society for Testing and Materials. Committee D-19 on Water. V. Series.
QH545.A17I47   1986   574.5'263   86-14076
ISBN 0-8031-0492-8

Copyright © by AMERICAN SOCIETY FOR TESTING AND MATERIALS 1986
Library of Congress Catalog Card Number: 86-14076

NOTE

The Society is not responsible, as a body,
for the statements and opinions
advanced in this publication.

Printed in Baltimore, MD
August 1986

Foreword

The symposium on the Impact of Acid Rain and Deposition on Aquatic Biological Systems was held in Bal Harbour, Florida on 29 Oct. 1984. The American Society for Testing and Materials' Committee D-19 sponsored the symposium. Billy G. Isom, Tennessee Valley Authority, and Sally D. Dennis and John M. Bates, Ecological Consultants, Inc., presided as symposium chairpersons; Sally D. Dennis and Eric L. Morgan, Tennessee Technological University, served as session chairpersons.

Related ASTM Publications

Validation and Predictability of Laboratory Methods for Assessing the Fate and Effects of Contaminants in Aquatic Ecosystems, STP 865 (1985), 04-865000-16

Aquatic Toxicology and Hazard Assessment (Seventh Symposium), STP 854 (1985), 04-854000-16

Ecological Assessment of Macrophyton: Collection, Use and Meaning of Data, STP 843 (1984), 04-843000-16

Ecological Assessments of Effluent Impacts on Communities of Indigenous Aquatic Organisms, STP 730 (1981), 04-730000-16

Aquatic Invertebrate Bioassays, STP 715 (1980), 04-715000-16

A Note of Appreciation to Reviewers

The quality of the papers that appear in this publication reflects not only the obvious efforts of the authors but also the unheralded, though essential, work of the reviewers. On behalf of ASTM we acknowledge with appreciation their dedication to high professional standards and their sacrifice of time and effort.

ASTM Committee on Publications

ASTM Editorial Staff

David D. Jones
Janet R. Schroeder
Kathleen A. Greene
Bill Benzing

Contents

Overview 1

Assessment of Aquatic Effects Due to Acid Deposition—
J. L. MALANCHUK, P. A. MUNDY, R. J. NESSE, AND D. A. BENNETT 4

A Preliminary Assessment of the Importance of Littoral and Benthic
Autotrophic Communities in Acidic Lakes—T. L. CRISMAN,
C. L. CLARKSON, A. E. KELLER, R. A. GARREN, AND
R. W. BIENERT, JR. 17

The Effects of Short-Term and Continuous Experimental
Acidification on Biomass and Productivity of Running Water
Periphytic Algae—L. PARENT, M. ALLARD, D. PLANAS, AND
G. MOREAU 28

Relationships of Spatial Gradients of Primary Production, Buffering
Capacity, and Hydrology in Turkey Lakes Watershed—
D. C. L. LAM, A. G. BOBBA, D. S. JEFFRIES, AND J. M. R. KELSO 42

Size-Dependent Sensitivity of Three Species of Stream Invertebrates
to pH Depression—J. W. ALLAN AND T. M. BURTON 54

Buffering Capacity of Soft-Water Lake Sediments in Florida—
T. E. PERRY, C. D. POLLMAN, AND P. L. BREZONIK 67

Linking Automated Biomonitoring to Remote Computer Platforms
with Satellite Data Retrieval in Acidified Streams—
E. L. MORGAN, K. W. EAGLESON, T. P. WEAVER, AND B. G. ISOM 84

Consideration of Total Ion Composition in Designing Toxicity Tests
Using Aluminum Salts and Mineral Acids—R. C. YOUNG 92

A Simple Method to Measure pH Accurately in Acid Rain
Samples—P. F. BOYLE, J. W. ROSS, J. C. SYNNOTT, AND C. L. JAMES   98

Author Index   107

Subject Index   109

Overview

There is considerable circumstantial evidence that implicates acid deposition effects on aquatic biological systems. However, investigations have not substantiated the existence of a single cause-effect relationship. Not all the suspected effects have been observed in apparently similar susceptible environments that have been studied, and significant differences between the results of apparently similar investigations of the same suspected effects have not been resolved. Great uncertainties are associated with the historical aquatic biological data base relevant to acid deposition effects. However, these uncertainties and the need for continuing research do not negate the present significance of acid deposition as one factor that appears in varying degrees to effect aquatic biological communities. This weakness in the historical data base has led to the present controversy about the contribution of acidification to depletion of biological diversity, especially with regard to fishery resources in ultraoligotrophic waters (water with low concentrations of dissolved ions). Lack of suitable biological (toxicological) and chemical methodologies appropriate to ultraoligotrophic waters have been factors contributing to the lack of hard data.

This volume brings together a diverse group of papers that have been used to link sources of acidification to aquatic biological effects. Air, water, soil (sediment), toxicology, and biological papers contribute to our understanding of lake/stream acidification processes and aquatic biological effects such as toxicity, productivity, and diversity. The volume opens with a paper by Malanchuk et al., who present the U.S. National Acid Precipitation Assessment Program (NAPAP) plan for 1985 and ongoing for ascertaining aquatic effects of acid deposition. Sections that describe the scope and organization of the assessment and a list of policy-related questions presented in the context of an Adirondack Mountain and New York regional case study are of particular interest. Fish are the principal biotic resource considered in the case study. Especially discussed are the uncertainties of: (1) the inferential process of relating acid deposition to loss of fish populations; (2) the application of laboratory-derived, dose-response functions to field situations; and (3) the correlating of chemistry to fish field surveys, some of which do not take factors into account such as fishing pressure, historical stocking practices, habitat changes, etc. that are necessary to identify the contribution of acid-deposition effects. The conceptual basis for an environmental model of reduced fishing benefits related to acidification is addressed also.

Crisman et al. and Parent et al. address the role of aquatic biological communities in acidified waters. Crisman et al. noted that while littoral and benthic communities are usually major contributors to autotrophic production in small lakes, they become much more important in acidified lakes where plankton are severely limited by lack of nutrients. Many studies have ignored this response in the past but should be included in future studies so that littoral-pelagic linkages can be quantified.

Parent et al. studied the effect of acidification on different trophic levels in freshwater microcosms. Perhaps surprisingly, at least to some, they found a contradiction to the hypothesis coupling acidification with the process of "oligotrophization" in that they found increased periphytic production following acidification.

Lam et al. investigated the underlying causes of acidification on the primary productivity of phytoplankton with the hypothesis that large regional differences in soil/sediment buffering capacity and watershed hydrology provide for large differences in growth rates even if acid deposition and nutrient conditions are the same. In their Turkey Lakes case study, they found that the pH, alkalinity, and dissolved inorganic carbon increased progressively downstream and were more important determinants of primary productivity than nutrients, sunlight, and temperature in these oligotrophic waters. The implication of their findings to the use and development of models relating acidification to eutrophication is also discussed.

The paper by Allan and Burton was more focused and reported size-dependent sensitivity of caddisflies, isopods, and snails to the effects of acidity in laboratory streams. Their recommendations were that biological programs designed to detect sensitivity to environmental stress should include testing of several life stages.

The paper by Perry et al. related the buffering capacity of soft-water lake sediments to artificial acidification in the laboratory and found in general that chemical- and biological-mediated processes in the sediments tend to counter impacts of acidity to the water column, which shows the importance of conducting highly integrated, interdisciplinary studies as suggested by Malanchuk.

Morgan et al. discussed linking automated biomonitoring (live organism responses) in acidified streams to remote computer platforms with satellite data retrieval. The advantage of this system is its capability to relate real-time organism responses to water quality during normal and unusual events such as spates or snowmelts. Another advantage is that it can be left at remote monitoring sites where real-time data is virtually impossible to obtain without automated sensing. The system described is significantly advanced over systems that have been used for a single waste stream or used in the laboratory.

Young's paper discusses identifying potential problems with the use of aluminum salts in laboratory toxicity tests of acidification effects, especially anion effects such as may be experienced with the use of aluminum chloride stock solutions, for example. There are many such chemical speciation problems in trying to separate acidification effects from other chemical/biological effects in poorly buffered waters. These problems include measuring pH, for example. It was soon found when acid deposition studies really got under way that conventional pH electrodes, which are designed to function in high conductivity solutions, simply did not work very well in very soft waters that scientists had to deal with. The paper by Boyle et al. addresses this problem and reported that increasing conductivity without changing the pH solved the measurement problem in combination with the use of the Ross pH electrode. Boyle et al. are sufficiently confident in their methods and in the electrode system to introduce them into the ASTM standards process for collaborative testing.

The collective papers in this volume will be of great help to those evaluating the impacts of acid deposition on aquatic biological systems and for evaluating mitigative benefits to such systems.

Billy G. Isom
Tennessee Valley Authority, Muscle Shoals, AL;
symposium chairperson and coeditor

John L. Malanchuk,[2] Patricia A. Mundy,[3] Ronald J. Nesse,[4] and David A. Bennett[5]

Assessment of Aquatic Effects Due to Acid Deposition[1]

REFERENCE: Malanchuk, J. L., Mundy, P. A., Nesse, R. J., and Bennett, D. A., "**Assessment of Aquatic Effects Due to Acid Deposition,**" *Impact of Acid Rain and Deposition on Aquatic Biological Systems, ASTM STP 928*, B. G. Isom, S. D. Dennis, and J. M. Bates, Eds., American Society for Testing and Materials, Philadelphia, 1986, pp. 4–16.

ABSTRACT: Increased concern over the impact of acid deposition on natural resources has caused the proliferation of substantial research in the area of effects. Often overlooked is the synthesis of this vast body of information into a coherent picture to be used for assessment and policy analysis. Relationships among research projects frequently are poorly defined or lacking, and problems of spatial and temporal resolution are abundant. The acquisition and use of historical data, for example, water quality and fish stocking data, to determine trends over time is problematical. Assessments of aquatic effects will be made in 1985, 1987, and 1989 under the National Acid Precipitation Assessment Program. A procedure is presented which attempts to organize existing information over space and time. Problems are highlighted and information needs made apparent.

KEY WORDS: acid deposition, assessment, aquatic effects, water chemistry

The question of what, if anything, to do about sulfur and nitrogen oxide emissions and acid rain has stirred intense scientific, technical, and political debate. The U.S. National Acid Precipitation Assessment Program (NAPAP) was implemented in 1980 to conduct research and assess the complex causes and effects of acid deposition. NAPAP has two primary objectives. First, it seeks to fill many gaps in the scientific understanding of the acid deposition phenomenon

---

[1] Although the research described in this article has been supported by the U.S. Environmental Protection Agency, it has not been subjected to Agency review and therefore does not necessarily reflect the views of the Agency and no official endorsement should be inferred.

[2] Aquatic effects team leader, Acid Deposition Assessment Staff, U.S. Environmental Protection Agency, Washington, DC 20460.

[3] Environmental protection specialist, Acid Deposition Assessment Staff, U.S. Environmental Protection Agency, Washington, DC 20460.

[4] Senior research economist, Acid Deposition Assessment Staff, U.S. Environmental Protection Agency, Washington, DC 20460, and Battelle Pacific Northwest Laboratory, Portland, OR 97232.

[5] Director, Acid Deposition Assessment Staff, U.S. Environmental Protection Agency, Washington, DC 20460.

through its comprehensive research program. Second, the program develops methods and performs assessments of the consequences of alternative acid deposition control strategies. The first of these assessments due in 1985 (the 1985 Assessment) focuses on the damages, to date, due to acid deposition.

The objective of this paper is to describe how the 1985 Assessment will deal with aquatic effects from acid deposition. The first section provides a brief description of the NAPAP organizational structure, the objectives of the 1985 Assessment and how it differs from subsequent assessments, and an overview of the aquatics portion of the assessment. Other sections describe the scope and organization of the assessment and provide a list of important policy-related questions presented in the context of a regional case study of the Adirondack Mountains of New York.

The NAPAP has ten working-level task groups, one for each of the national program's nine research categories and one for international activities. These technical groups include program managers and experts from the participating federal agencies, National Laboratories, and elsewhere. They are responsible for detailed planning and conducting of research in their assigned areas. Figure 1 shows the task force's operating structure and identifies the task groups and the agencies responsible for their leadership.

Most of the task groups shown in Fig. 1 focus on the advancement of scientific information on the nature, causes, effects, and controls of acid deposition. Their activities are centered on the setting of research priorities, the gathering and analysis of data, study of various physical and ecological processes, and the development of predictive models. For example, the Aquatic Effects Task Group (Task Group E) has responsibility for conducting research on acidification of lakes and streams and the attendant ecological effects. Only the Assessments Task Group (Task Group I) is directed toward integrating the research results, focusing the research upon public policy concerns, and assessing the potential results of various measures to reduce acid deposition.

The 1985 Assessment seeks to meet a set of NAPAP objectives, in particular three specific objectives related to aquatic effects. They are to:

1. Quantify aquatic resources and estimate, with attendant uncertainties, the environmental and economic damages to date caused by acid deposition.
2. Within the bounds of error, estimate the rate at which acidification is occurring.
3. Assess the costs and effectiveness of aquatic mitigation techniques.

Organization and Scope of the Aquatic Assessment

With respect to aquatic systems, we will discuss qualitatively the possible linkages from emissions to the economic impact of changes in the resource base and, in part, will attempt to quantify known links of the mechanistic pathways and any observed changes over time. Later assessments, in 1987 and 1989, will

ORGANIZATION CHART OF THE NATIONAL ACID PRECIPITATION ASSESSMENT PROGRAM

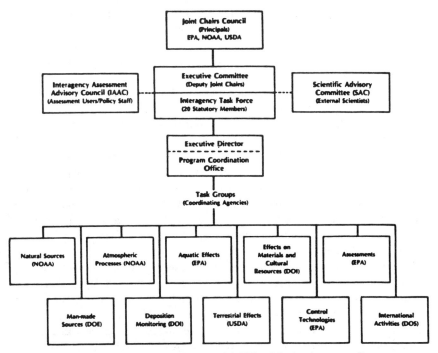

FIG. 1—*Organization chart of the National Acid Precipitation Assessment Program.*

have as their main objective the quantitative fully integrated assessment. It is our opinion that, at this time, the uncertainty of the information is so great and the pathways from emissions to effects so poorly quantified that it would be misleading to attempt any quantitative, fully integrated assessment in 1985.

Fish are the principal biotic resource considered; studies of other flora and fauna will be reported but are not analyzed quantitatively. Many uncertainties exist in the inferential process of relating acid deposition to the loss of fish populations in lakes and streams. One such uncertainty is the application of laboratory-derived, dose-response functions to field situations to predict the percent survival of fish to a dose of acid or acid in combination with other potential toxicants. Another uncertainty in correlating chemistry to natural fish populations is the use of fish field surveys themselves. Some surveys do not take into account factors such as fishing pressure, stocking practices, changes in habitat, etc., necessary to separate the effects on sport fish due solely to acidity or acid-derived substances.

The 1985 Assessment also will include estimates of the effects of acid deposition on fish in terms of economic damages, such as changes in well-being from reduced fishing opportunities in the Adirondack Mountains of New York. The economic analysis will attempt to link quantitatively the change in fishable hectares to values reflecting perceptions of the importance of changes in recreational fishing.

National Assessment

Although the 1985 Assessment will analyze data for an intensive case study of the Adirondacks and for additional regional case studies of the Upper Midwest, the Blue Ridge Province, and Pennsylvania, and to a lesser extent for a multistate analysis of New York and New England, the national level analysis will be limited to reporting available water resource data and analysis of aquatic chemistry data from the 1984 National Surface Water Survey (NSWS) [1].

Available data on the extent of physical resources on, for example, hectares of lakes and kilometers of streams, classified by size, and data on geology, land use, and vegetation will be presented in preparation for the 1987 national, integrated assessment. A national map of surface water alkalinity [2] depicting the distribution of alkalinity by several categories will be used to identify areas of low alkalinity water. Several regional alkalinity maps, at finer resolution, also will be presented and discussed [3].

The NSWS was designed to collect statistically representative samples for water quality analysis for regions with substantial numbers of water bodies of low alkalinity for the nation [1]. It is envisioned as a three-phase program. Phase I is to determine the extent of low alkalinity lakes and streams and their associated chemistry (Table 1). Phase II will quantify the biological components of the system and resolve spatial and temporal variability of chemical parameters for selected water bodies. Phase III consists of the long-term monitoring of regionally representative lakes to quantify and document future changes in the chemistry and biology of aquatic systems. The NSWS Phase I lake data will be included in the 1985 Assessment for the Northeast, Southeast, and Upper Midwest. Together with the national and regional alkalinity maps, the NSWS will provide a reasonably accurate estimate of the chemical status of the national aquatic resource relevant to acid deposition. Although this information alone is insufficient to answer any questions about trends or rates of acidification, it does provide the first statistically valid national scale data on the chemical status of lakes. In many instances, these data will provide the basis for future answers to questions concerning acidification trends.

Estimates of changes in water chemistry will be presented based upon analyses of current chemical data collected in the NSWS and historical chemical data collected from a variety of sources including the Acidification Chemistry Information Database (ACID) [4]. Changes in water chemistry will be related to surrounding watershed attributes, and, insofar as possible, watersheds will be categorized into different response types. This activity assumes a causal linkage among watershed characteristics and the water chemistry of the receiving water, but only correlative analyses are planned.

TABLE 1—*Water quality parameters measured in the National Surface Water Survey.*

Parameter
Acidity
Acid neutralizing capacity
Aluminum, total
Aluminum, extractable
Ammonium, dissolved
Calcium, dissolved
Chloride, dissolved
Fluoride, dissolved—total
Inorganic carbon, dissolved
Iron, dissolved
Magnesium, dissolved
Manganese, dissolved
Nitrate, dissolved
Organic carbon, dissolved
pH
Potassium, dissolved
Silica, dissolved
Sodium, dissolved
Sulfate, dissolved
Specific conductance
Total phosphorus

Multistate Assessment

The multistate analysis of New York and the seven New England states will rely heavily on the regional alkalinity map [2], the NSWS [3], and the ACID project [4]. These projects will provide the bulk of the chemical information for the region. The ACID is a large data base derived from the Environmental Protection Agency (EPA STOrage and RETrieval (STORET) system and consists of stations which met a criterion of at least ten observations with at least one observation in each of five separate years. The data base was supplemented with data not resident in the STORET system.

A study published by the U.S. Geological Survey [5] reports the data and trends for stations in the multistate region. Both the Smith and Alexander study [5] and the Hendrey et al. study [4] independently suggest that surface water sulfate concentrations in the Northeast are decreasing and that this observation is related to reduced emissions in the same region.

The National Academy of Science (NAS) also is conducting a study on trends in acid deposition. This NAS committee has identified a series of sampling sites which have "high quality" data, many of which are in the multistate region. This analysis is not yet completed, but it will be included in the final assessment document.

The studies supporting the biological portion of the multistate analysis consist of surveys of organisms in lakes of Maine, New Hampshire, Vermont, Rhode Island, and a few observations in Massachusetts. These data are being merged into another data base called the Fish Information Network [6]. This will facilitate the study of associations between fish status and water chemistry. These data will be utilized in answering a series of questions about acid deposition that provide the focus for this assessment and the regional case studies. The questions, however, are not presented here; they are stated and discussed in the Adirondack regional case study that follows.

Adirondack Regional Case Study

Much of the evidence suggesting that acidification of water bodies has occurred with consequent loss of biotic resources comes from the Adirondack Mountains of New York State. These observations coupled with the fact that data are sparse or lacking for all but a few other regions led to the decision to focus the 1985 Assessment activity on four regional case studies, the Adirondacks study being the most detailed. Following are several questions and, for some, the sources of information contributing to the answers. If the information is not presently known but is forthcoming, that is stated and the procedure to be used to answer the question is given. Finally, if an important question cannot be answered, an area of future research is identified. It should be noted that throughout this paper the word "acidification" is used to describe the process by which aquatic systems experience a decrease of pH to less than 5.6.

Have There Been Changes in the Chemical Status of Surface Waters?

Yes. We believe that perhaps as many as 200 lakes in the Adirondacks have lost alkalinity since the 1930s and have experienced an increase in acidity. Moreover, many low alkalinity waters are considered sensitive to acidification should deposition increase. Our analysis will employ historical and current alkalinity and sulfate data to answer the question. Stations that have observations over time will be especially useful assuming that problems of changes in analytical methodologies over the period of record can be rectified.

If Acidification Has Occurred, Was it Predominantly Natural or Anthropogenic in Origin?

It may be possible to examine data from lakes for correlations that suggest whether a lake became acidified by organic acids or other means. Causation cannot be demonstrated and is not implied in Fig. 2, which illustrates the procedure to infer the means of acidification by organic acids. Certainly the presence of sphagnum mats or other acidophilic vegetation, high dissolved organic carbon, or highly colored water suggests that acidification by organic acids may be important in a particular system's chemistry. The contribution of strong acid anions from natural processes such as vegetative growth, succession, nitrification, sulfur oxidation, podzolization, etc. have not been considered. Certainly these processes may be a contribution to the strong acid anion pool, but the relative importance

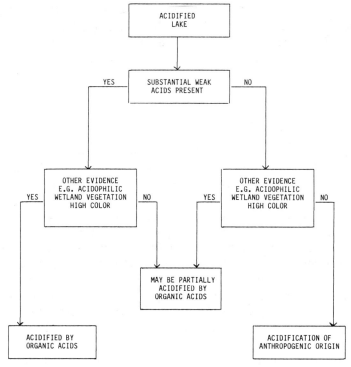

FIG. 2—*Flowchart to determine whether acidification is from organic acids or anthropogenic in origin.*

of these processes versus anthropogenic inputs has not been addressed. Later assessments will consider these issues.

What Watershed Characteristics are Important in Determining Sensitivity to Acidification? What Role is Played by Soils, Hydrology, Weathering Rates, Vegetation Types, Fires, Logging, Recreational or Second Home Development, Beaver Activity, etc., in Translating Deposition Chemistry into Surface Water Chemistry?

No direct causal linkage to deposition will be made. Rather, we will examine correlations among various land use/watershed parameters and surface water chemistry. Previous attempts have been made to correlate various parameters to water chemistry. Eilers et al. [7] performed cluster analysis on 275 Wisconsin lakes and defined three primary lake types: precipitation-dominated, groundwater-dominated, and surface runoff-dominated. They found that lakes in the precipitation-dominated class are very different from other lakes in that they lack sources of groundwater and watershed buffering inputs. These lakes would not be expected to be sensitive to short-term episodes. Therefore, hydrology and watershed area must be considered in any assessment of lake sensitivity to long-term deposition.

A similar approach to that of Eilers et al. was employed in a study of lakes in Upper Michigan but met with more limited success [8].

This statistical approach has been extended by combining it with a deterministic model [9]. Although this work is still in progress, the basic concept is the association of a set of model coefficients with each lake cluster followed by performing simulations which allow each lake to be subjected to a desired loading scenario to predict the lake's equilibrium alkalinity. We agree that many factors influence water chemistry and feel that previous attempts that met with limited success in some regions may have been due partially to poor resolution of the data set. We are in the process of collecting some data at a finer level of resolution; in some cases data exist for one-acre grid cells. We expect that improved resolution of the data will provide improved statistical relationships. A series of maps will be prepared to overlay detailed soils, historical data on fires and logging, development, beaver activity, and other watershed information with water chemistry. If maps are not feasible, tables of information will be used to compare soils, geology, logging, fires, vegetation type, etc. to surface water chemistry.

Several speculative hypotheses have appeared in the literature and have yet to be evaluated. Examples of these are:

− 1. Fires or logging in the Adirondacks in the early 1900s contributed base cations to lake systems and allowed previously fishless lakes to support sport fish populations. These systems are now returning to their original acid state, perhaps at an accelerated rate, because of effective forest fire prevention [10].

− 2. Wildlife management decisions have been responsible for the resurgence of beaver in the Adirondacks. Increased beaver activity, particularly dams across small, high alkalinity inlets to acidified lakes, causes fish declines by preventing fish from taking refuge in these streams during episodes of low pH such as during snowmelt [11].

These hypotheses will be investigated by comparing regions of fire, logging, and beaver activity with their respective lake chemistry. The Adirondacks case study offers a unique opportunity for this analysis.

What is the Extent of Low pH, Low Alkalinity Lakes?

The extent of acidic waters will be examined by at least two different means. First, the impacted resource is compared to the total resource. In this case, we will know the total number and acreage of lakes in the Adirondacks. Also, it is assumed that all acidic lakes have been identified. Thus, the number (or acreage) of lakes that fall into a certain pH or alkalinity class relative to the total resource can be specified. The alkalinity classes to be used are <100, 100 to 200, and >200 µeq/L. Second, the low pH, low alkalinity resource will be compared to the total resource by size class of lakes. It would be important to know whether small lakes become acidic preferentially. For example, if only a small areal extent of lakes is acidic but consists entirely of small lakes, then one could argue that the entire small lake resource is acidic [12].

What is the Rate of Acidification?

For a limited number of lakes, sufficient data exist for a period of record that allows the determination of trends. The time intervals of relatively constant deposition, or those where changes were at least unidirectional, will be examined. It is not likely that sufficient data will be available to allow extrapolation to the entire aquatic resource. We hope to be able to accomplish this for the 1987 assessment.

Recently, a major research initiative was established to investigate the response time of watersheds subjected to acid deposition and the mechanisms that control the response rate. Termed the "direct/delayed response" initiative, it will study the terrestrial portion of the watershed's ability to neutralize atmospheric acidity as water passes through the system and whether the ability to neutralize acids changes over time. Within-lake processes will not be evaluated by this initiative. Four basic components appear to be central to the question: extent and duration of contact, mineral weathering replacement of base cations, anion retention, and ion exchange buffering.

What are the Effects of Acidification in the Adirondacks?

Have there been changes in lake chemistry, fish populations, and habitat or in other aquatic flora and fauna? Cores have been used to examine the changes of metal content of lake sediments over time. Norton et al. [13] have speculated that acidification of lakes may be responsible for mobilizing sediment-bound metals, thereby increasing concern for the bioaccumulation of these materials and possible entry into human food chains. Another investigator has used sediment cores to examine changes in diatom and chrysophyte assemblages [14,15] over centuries, relating this information to historical changes in pH. Sediment core data cannot be extrapolated easily to the entire resource because these data exist only for a limited number of lakes. Any conclusions drawn from these data will be bounded by discussion of their inherent uncertainty, which arises from the mixing and leaching of sediment layers and the lack of means of assessing causality.

With respect to fish habitat, there may have been changes in the water chemistry that prevent fish from living in areas previously inhabited. Care will be taken to differentiate between actual and potential fish habitat and the changes that have occurred in each. It is important to know whether fish ever existed in a lake in order to correctly assess acid deposition effects.

There are scattered reports in the literature concerning impacts of acidification on Adirondack flora and fauna other than fish [15,16]. These reports cannot be used directly in the analysis because they are not comprehensive. However, key findings will be noted and discussed on a case-by-case basis.

Fish will be discussed by species, emphasizing recreational species, and will draw heavily on the work of Baker [1]. Changes in fish populations will concentrate on presence/absence information and will account for changes in fish species composition over time. This will be done on a lake-by-lake basis. Tabulations of species changes for each lake will be presented. In addition, general patterns will be sought to determine whether any common trends exist. This will provide increased confidence in the analysis since in many cases interpretation is confounded by changes in stocking practices and fishing pressure. Any quantitative treatment of the uncertainty of this information will be difficult. In any event, uncertainty will be handled as completely as possible.

A final point to be considered is whether similar changes have occurred in nonacidic systems. If such changes have occurred, they will be noted and compared to the impact in acidic systems.

What is the Extent of the Acidification Impact? More Specifically, Is It Possible to Quantify Actual/Potential Fishable Hectares Lost?

This question supplies the information needed to produce the economic analysis of fishing damages. We propose three possible approaches to estimate the change in fishable hectares. The first method is to sample directly fish presence and absence (present and historical) for all lakes that have data and extrapolate to the total number of lakes in the Adirondacks. Actual fish presence and absence, by species, for a small number (less than 200) of Adirondack lakes is known. Historical information on presence and absence also is available. If there are sufficient numbers with both historical and contemporary data, the changes can be extrapolated to the entire Adirondack region. Lacking a sufficient number with both sets of data, each data base can be individually extrapolated to the entire Adirondacks. After investigating changes in presence/absence due to stocking practices (including adjustments for changes in water chemistry) or other lake-specific influences, the difference in total fishable hectares would be assumed to be due to acid deposition.

The second method (Fig. 3) is to employ an empirical model that links water chemistry to fish presence/absence and then extrapolate to the total resource. Data for fish presence/absence and water chemistry are being used to develop a model that, using historical data on water chemistry, will allow predictions of whether a particular lake was able to sustain fish. In addition, the model can be used to predict fish presence/absence in contemporary Adirondack lakes that have known water chemistry. The predictions would need to be extrapolated to obtain estimates of changes in fishable hectares for the entire Adirondack region.

The third proposed method would be to predict historic lake chemistry followed by application of the fish presence/absence model. Models relating present lake chemistry to variables such as watershed weathering could be "run backwards" using known present lake chemistry and weathering rates to "hindcast" historical lake chemistry. The historical lake chemistry could then be used in a fish presence/

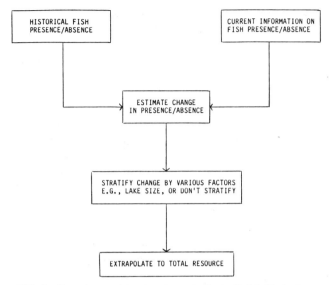

FIG. 3—*Procedure to determine change in "actual" fishable hectares.*

absence model. Again, extrapolation to all lakes would be necessary to compare changes in fishable hectares. Each of these procedures may be subject to substantial errors and uncertainties. Many of these uncertainties, for example weathering rates, are difficult to quantify. These potential errors will be discussed in the context of these estimates.

What is the Economic Value of the Reduced Fishing Opportunities?

The 1985 Assessment also will include estimates of the effects of acid deposition on fish in terms of economic damages such as changes in well-being from reduced fishing opportunities in the Adirondack Mountains of New York. The section on fishing will include changes in the level of consumer surplus associated with reduced fishing in the Adirondacks. Consumer surplus is an economic concept used to measure changes in well-being. Since there are large uncertainties in the level of physical effects (changes in fishable hectares or daily catch rate), a range of values will be displayed. A second portion of the economic analysis will evaluate the magnitude of "nonuser" or "intrinsic" values for changed environmental or recreational quality. For example, individuals may value the existence of an environmental improvement even though they do not expect to use the affected resource, that is, nonusers may value environmental resources they hope to protect for future generations. This paper will discuss only lost recreational fishing opportunities.

Several alternative methods for calculating regional estimates of the economic effects of reduced fishing opportunities were considered. The valuation of regional effects is different from the economic studies in the past that focused on the value of a specific recreational site. The effects of acid deposition over a whole region such as the Adirondacks make a disaggregated site-by-site approach intractable. As a result, some methods development was necessary to devise appropriate models to value a reduction in recreational opportunities throughout a region. It is also important to distinguish between the values from fishing and those from other recreational benefits associated with fishing, such as enjoying the scenery.

The economic product for the 1985 Assessment will be an analysis using a statistical travel-cost model [17] based on Adirondack historical data that relates a measure of economic value (for example, a consumer surplus measure) to the characteristics of each fishing site. If successful, the model will allow an assessment of the economic damages, expressed in dollars, to recreational fishing from acid deposition. The data on effects needed to perform the calculation would be the levels of the site characteristics, such as fishable hectares that exist in the absence of any acid deposition from manmade sources. Estimates of future damages could be made by forecasting future values of these site characteristics under different acid deposition scenarios. However, the emphasis of the 1985 Assessment is on obtaining an estimate of current annual damages based on the current rate of deposition.

An alternative (or perhaps an addition) to the travel-cost calculations is the use of a model linking acid deposition and changes in participation [18]. A participation model links an individual's decision to fish at a particular site given the site's characteristics. In this approach, changing the level of fishable hectares changes the number of fishing days. The estimates of changes in participation can be linked to an imputed unit value for fishing days from other studies to obtain an estimate of regional damages. There are several potential sources for imputed values, including surveys on the value of fishing opportunities and other studies of fishing values.

Estimating nonuse values will present a number of difficult conceptual and empirical problems. For example, how do you include the values of the many persons physically distant from where acid deposition effects are occurring into a regional or national estimate of nonuse values? These problems are being considered in an ongoing study of how to design a survey capable of answering the valuation questions. However, the development, testing, and refinement of the survey will not be completed in time for the 1985 Assessment. A qualitative discussion with some possible bounds on the magnitude of potential damages is anticipated for the 1985 Assessment.

References

[1] Office of Research and Development, "National Surface Water Survey, Phase I, Research Plan," U.S. Environmental Protection Agency, NAPAP Project E1-23, Washington, DC, 1984.
[2] Omernik, J. M. and Powers, C. F., "Total Alkalinity of Surface Waters—A National Map," EPA Report EPA-600/D-82-333, U.S. Environmental Protection Agency, Corvallis, OR, 1983.
[3] Omernik, J. M. and Powers, C. F., "Total Alkalinity of Surface Waters—Regional Maps," U.S. Environmental Protection Agency, Washington, DC, 1984.

[4] Hendrey, G. R., Hoogendyk, C. G., and N. F. Gmur, "Analysis of Trends in the Chemistry of Surface Waters of the United States," Informal Report BNL 34956, Vol. 1, Brookhaven National Laboratory, Upton, NY, May 1984.

[5] Smith, R. A. and Alexander, R. B., "Evidence for Acid-Precipitation-Induced Trends in Stream Chemistry at Hydrologic Bench-Mark Stations," Geological Survey Circular 910, U.S. Geological Survey, Alexandria, VA, 1983.

[6] Baker, J. P., "National Acid Precipitation Assessment Program Effects Research Review: Research Summaries," North Carolina State University, Raleigh, NC, 21–25 Feb. 1983.

[7] Eilers, J. M., Glass, G. E., Webster, K. E. and Rogalla, J. A., *Canadian Journal of Fisheries and Aquatic Sciences,* Vol. 40, 1983, pp. 1896–1904.

[8] Schnoor, J. L. and Nikolaidis, N. P., "Factors Controlling Water Chemistry and Sensitivity to Acidification of Lakes in Upper Michigan," report, U.S. Environmental Protection Agency, Duluth, MN, Oct. 1983.

[9] Schnoor, J. L., Palmer, W. D., Jr., and Glass, G. E. in *Modeling of Total Acid Precipitation Impacts,* Chapter 7, J. L. Schnoor, Ed., Butterworth Publishers, Boston, MA, 1984, pp. 155–174.

[10] Retzsch, W. C., Everett, A. G., Duhaime, P. F., and Nothwanger, R., "Alternative Explanations for Aquatic Ecosystems Effects Attributed to Acidic Deposition," report, The Utility Air Regulatory Group, c/o Hunton and Williams, Washington, DC, March 1982.

[11] Duhaime, P. F., Everett, A. G., and Retzsch, W. C., "Adirondack Land Use: A Commentary on Past and Present Impacts on Terrestrial and Aquatic Ecosystems," report, The American Petroleum Institute, Washington, DC, Feb. 1983.

[12] Pfeiffer, M. H. and Festa, P. J., "Acidity Status of Lakes in the Adirondack Region of New York in Relation to Fish Resources," report, New York State Department of Environmental Conservation, Albany, NY, Aug. 1980.

[13] Norton, S. A., Williams, J. S., Hanson, D. W., and Galloway, J. N., "Changing pH and Metal Levels in Streams and Lakes in the Eastern United States Caused by Acidic Precipitation" in *Restoration of Lakes and Inland Waters,* EPA Report 440/5-81-010, Environmental Protection Agency, Washington, DC, 1981, pp. 446–452.

[14] Charles, D., *Proceedings of the International Association of Theoretical and Applied Limnology,* Vol. 2, in press.

[15] Smol, J., Charles, D. F., Whitehead, D. R., *Nature,* Vol. 307, 1984, pp. 628–630.

[16] Jacobson, J. S., Ed., *Proceedings,* Second New York State Symposium on Acid Deposition, Center for Environmental Research, Albany, NY, 1983.

[17] Burt, O. F. and Brewes, D., "Estimation of Net Social Benefits from Outdoor Recreation," *Econometrica,* Vol. 39, 1971, pp. 813–827.

[18] Vaughn, W. J. and Russell, C. S., "Freshwater Recreational Fishing," Resources for the Future, Washington, DC, 1982.

Thomas L. Crisman,[1] Chandler L. Clarkson,[2] Anne E. Keller,[2] Robert A. Garren,[2] and Raymond W. Bienert, Jr.[2]

A Preliminary Assessment of the Importance of Littoral and Benthic Autotrophic Communities in Acidic Lakes

REFERENCE: Crisman, T. L., Clarkson, C. L., Keller, A. E., Garren, R. A., and Bienert, R. W., Jr., "**A Preliminary Assessment of the Importance of Littoral and Benthic Autotrophic Communities in Acidic Lakes,**" *Impact of Acid Rain and Deposition on Aquatic Biological Systems, ASTM STP 928,* B. G. Isom, S. D. Dennis, and J. M. Bates, Eds., American Society for Testing and Materials, Philadelphia, pp. 17–27.

ABSTRACT: Most of the research examining changes in aquatic biota relative to increasing acidity has concentrated on pelagic components. The possible influence of littoral-pelagic interactions on observed relationships has largely been ignored, and thus interpretation of pelagic biotic data unfortunately often is taken out of context of the whole ecosystem. While littoral and benthic communities are major contributors to the autotrophic production of small lakes in general, it is likely that their share assumes greater importance in acidic lakes where phytoplankton are often severely nutrient-limited. The availability of nutrients in the pelagic zone of acidic lakes may be regulated by littoral and benthic processes. Available data suggest that while the structure of littoral and benthic autotrophic communities has a direct influence on benthic invertebrates, it exerts an indirect control on zooplankton and fish principally through the quality and quantity of habitat and food resources. Future investigations of biota in acidic soft-water lakes should concentrate on quantifying such littoral-pelagic linkages.

KEY WORDS: acid rain, limnology, trophic-level interactions, littoral zone, lakes, zooplankton, macrophytes, bacteria, phytoplankton, algae, primary production, benthic invertebrates, fish

Pelagic Community Responses to Increasing Acidity

Acid deposition aroused scientific interest only during the 1960s and 1970s when researchers were able to link the observed historical demise of fish popu-

---

[1] Associate professor, Department of Environmental Engineering Sciences, University of Florida, Gainesville, FL 32611.

[2] Graduate research assistant, Department of Environmental Engineering Sciences, University of Florida, Gainesville, FL 32611.

lations in Scandinavian lakes to progressive acidification from anthropogenic sulfur loading to the atmosphere. Because few historical data exist for individual lakes spanning the pre/post acidification period, scientists have relied on biotic surveys of large numbers of lakes comprising a gradient of pH to predict how individual biotic components of lake ecosystems likely will respond to progressive lake acidification.

A great deal of information exists on plankton and fish responses to acidification. Phytoplankton species richness and diversity decrease with decreasing pH with the greatest changes generally occurring in lakes with pH < 5.6 [1-3]. With the exception of the study conducted by Kwiatkowski and Roff [4], most investigations have noted that the importance of blue-green algae is greatly reduced in acidic lakes. Although several synoptic surveys have reported trends of lower phytoplankton abundance with decreasing pH [1,4], additional workers indicate that acidification of temperate lakes is not necessarily accompanied by a significant reduction in algal biomass [5-6]. Several investigators have suggested that phytoplankton biomass is controlled more by phosphorus concentrations than by pH. This hypothesis is supported by results of enclosure and whole-lake experiments in which acidic water has been enriched with phosphorus [8-11].

Zooplankton species richness, abundance, and biomass generally decrease with decreasing pH [12-14]. Small species usually dominate the macrozooplankton assemblages of North American acidic lakes [14], and rotifers often assume numerical abundance over macrozooplankton [15]. Although largely ignored as a zooplankton component, a preliminary investigation of ciliate protozoans in Florida lakes suggests that while ciliate biomass is generally reduced in acid lakes [16], trophic state rather than pH appears to be the controlling factor [17]. A similar positive correlation between both total zooplankton abundance [18] and biomass [13] and chlorophyll a in acidic lakes has been noted. These results coupled with the pronounced response of zooplankton to the fertilization of acid lakes [11] indicate that autotrophic productivity can be as important as pH in structuring zooplankton communities in acidic lakes.

Based on intensive surveys of lakes representing a gradient of pH conditions, a major reduction in both the species richness and biomass of fish communities has been noted with increasing acidity in Norway [19], Sweden [20], Canada [21], the northern United States [22], and Florida [23]. In all of the temperate lake regions, there is a progressive loss of individual fish species with increasing acidity, while in subtropical Florida lakes representing a comparable pH gradient, there is no elimination of individual species. While direct acute and chronic pH effects may be expected, indirect effects of lowered pH on the availability of sodium, calcium [2], aluminum [24], and heavy metals [25] have been implicated as the primary reason for the loss of fish populations in acidic waters.

Of all the biotic communities in the pelagic zone, the response of bacterioplankton to lake acidity is the least clear. A number of investigators [26-29] have reported that bacteria abundance is not significantly reduced in acidic (<5 to 6 pH) temperate lakes, while others working in the same geographic areas [30–32] and subtropical Florida [33] have found significantly reduced numbers. The abundance of such conflicting data suggests that additional factors such as organic substrate availability are equally as important as pH in the observed relationship.

Benthic Community Responses to Increasing Acidity

In marked contrast to the pelagic zone, our understanding of the response of benthic (both littoral and profundal) communities and processes to acidification is incomplete. Both the species richness and standing crop of benthic invertebrate communities decline with increasing acidity [5,34–35], but the effect of acidification on benthic secondary productivity is unknown [6]. The greatest sensitivity to low pH is displayed by molluscs, macrocrustaceans, plecopterans, and ephemeropterans [36]. While such changes in benthic communities relate to assemblages living either on macrophytes or at the sediment surface (epifauna), recent evidence suggests that species living in the sediments (infauna) are not significantly affected by lake pH [37].

With acidification, submergent macrophyte communities in temperate lakes often become simplified through the loss of most angiosperm species and a shift to dominance by the bryophyte *Sphagnum* [7,38–39] or *Juncus* [40–41]. Recent studies suggest that the structure of autotrophic communities of the littoral zone may be further altered through an expansion of epiphytic algal biomass [28,42] or benthic algal mats [43–46].

Grahn et al. [47] reported that a shift from bacteria to fungi as the dominant decomposers in recently acidified Swedish lakes has resulted in decreased decomposition rates and increased accumulation of coarse organic detritus. The principal evidence used by them and others [7,38] for this shift was the presence of felt-like fungal mats in littoral and profundal areas of many acidic lakes. Lazarek [43] examined such felt-like mats from acidic Swedish lakes using scanning electron microscopy and found that, rather than being fungal, they were composed principally of filamentous blue-green algae, including *Lyngbya*, *Oscillatoria*, and *Pseudoanabaena* covered with a film of partially decomposed organic matter. This conclusion was supported by Hendrey and Vertucci [39]. In summary, the effect of acidification on benthic decomposition rates is not clear, as some authors [7,19,48] reported a reduction while others [49–50] failed to show a difference relative to less acidic lakes.

As is generally the case in limnology, most of the research on soft-water acidic lakes has concentrated on pelagic communities and processes while largely ignoring the littoral and profundal zones. In part, this reflects the fact that pelagic communities are much easier to quantify. Hypotheses generated regarding the response of various pelagic communities to acidification are usually taken out of context of the whole system and ignore the influences of littoral-pelagic interactions on observed relationships.

Influence of Littoral Vegetation on Total Ecosystem Autotrophic Production

The contribution of the littoral zone to total lake ecosystem primary production is poorly known. Wetzel [51] noted that most lakes are small in volume and area. Such lakes have a morphometry that maximizes the potential contribution of the littoral zone to total system production relative to the contribution of the pelagic zone based on potentially colonizable area for benthic autotrophs. This certainly applies to most acidic lakes, which tend to be small. The contribution of littoral and benthic communities to total lake production has been estimated to be extremely high in Czechoslovakian fish ponds (93%), Borax Lake (50%), and Lawrence Lake (75%) in spite of the limited areal extent of these communities (summarized in Wetzel [51]). Recently, Canfield et al. [52] demonstrated that the most valid estimate of lake trophic state includes both pelagic chlorophyll a concentrations and the percent of total lake volume occupied by macrophytes.

As discussed earlier, phytoplankton biomass decreases with increasing acidity in temperate lakes, but the magnitude of the response is also strongly influenced by the availability of phosphorus. Within the vegetated littoral zone, there is a general reduction in both the species richness and biomass of submergent macrophyte communities, coupled with increased importance of epiphytic algae, benthic algal mats, and *Sphagnum*.

In clear-water subtropical Florida lakes representing a comparable gradient of acidity as studied in the temperate zone, phytoplankton biomass is generally lower in acidic lakes [18], but the littoral and benthic community response to acidification differs somewhat from that reported in the temperate zone. In part, this may reflect the absence of *Sphagnum* in peninsular Florida (Fig. 1).

Two general patterns of littoral zone community structure are exhibited by acidic (pH 4 to 5.5) subtropical Florida lakes. In many lakes, submergent macrophytes are totally absent, and the littoral zone has been contracted to a narrow fringe of emergent grasses. Epiphytic algal biomass appears to be extremely high in the grass zone, and, although algal mats do not develop in deeper areas devoid of macrophytes, benthic algae are numerous in the interstitial spaces within the sandy sediments. At one site, Lake Sheelar (pH 4.90), we have estimated that the midsummer biomass of emergent macrophytes (691 kg organic carbon) is 36 times greater than that of phytoplankton (19.3 kg organic carbon). These data provide a highly conservative estimate of the importance of the littoral and benthic communities to total production in acidic systems as we have not assessed the contribution of either epiphytic or benthic algae.

Other comparably acidic lakes, most notably in the Ocala National Forest of central Florida, fail to show a reduction in either the species richness or extent of the submergent macrophyte community along a gradient of declining pH. Submergent vegetation in such lakes frequently is 2 to 3 m high and dominated by *Utricularia* and *Websteria*. Epiphytic algae are very lush, and during summer epiphytic algal "clouds" in excess of 2 m in diameter are found. As with the

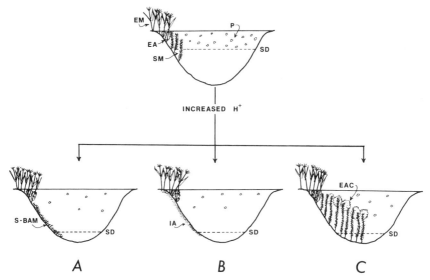

FIG. 1—*Frequently observed structural changes in lacustrine autotrophic communities with increasing acidity: (A) shift in littoral and benthic communities to favor* Sphagnum *and benthic algal mats (S-BAM) as often seen in temperate lakes; (B) increased importance of interstitial algae (IA) observed in sand bottom Florida lakes; and (C) expansion of submergent macrophytes (SM) and appearance of epiphytic algal clouds (EAC) in other Florida lakes. Note that phytoplankton (P) is reduced, epiphytic algae (EA) and Secchi disk transparency (SD) increase, and emergent macrophytes (EM) remain unchanged in all three acidic lake types.*

other group of acidic lakes in Florida, pelagic chlorophyll *a* concentrations rarely exceed 5 mg/m^3.

Although detailed quantitative data are lacking, it appears that littoral and benthic communities assume greater importance as contributors to total ecosystem autotrophic production with increasing acidity. Even in lakes with littoral zones restricted to a fringe of emergents such as described for Lake Sheelar, Florida, littoral zone production is likely of prime importance. Secchi disk transparency generally increases with lake acidification due to a combination of factors, including reduced phytoplankton biomass, dissolution of iron and manganese colloids, and precipitation of suspended organics by aluminum [6,53]. Thus, for individual lakes undergoing cultural acidification, as phytoplankton biomass declines, the contribution of littoral and benthic communities to total ecosystem autotrophic production actually could increase as both communities expand into deeper water as a result of an increase in the depth of the photic zone.

Declining biomass of phytoplankton or submergent macrophytes or both is often used to support the idea that autotrophic production is greatly reduced in acidic lakes [36]. Such interpretations should be approached cautiously. One cannot assume that if pelagic production declines then littoral and benthic production will behave similarly. Even within the littoral zone, declining submergent macrophyte biomass often coincides with an expansion in epiphytic algae, benthic algal mats, and, frequently in the temperate zone, the bryophyte *Sphagnum*.

Wetzel [*51*] noted that emergent macrophytes are perhaps the most productive macrophyte community and that submergents are among the most unproductive, having similar productivity to phytoplankton on an areal basis. In marked contrast, the productivity of epiphytic algae is usually much greater than submergents. Unfortunately, productivity data for benthic algal mats are generally lacking. Thus, although phytoplankton and submergent macrophyte biomass may be reduced, total ecosystem autotrophic production in acidic lakes may be unaltered as system dominance shifts to highly productive epiphytic and benthic algal communities.

Finally, the littoral zone can directly affect ecosystem productivity through its influence on water chemistry. The bulk of evidence suggests that submergent angiosperms rely on sediment interstitial water for most of their phosphorus needs [*54–55*]. Submergent macrophytes, in turn, release much of the phosphorus obtained from sediment sources through active nutrient pumping and decay. Export of such phosphorus from the littoral to the pelagic zone for utilization by phytoplankton is controlled by the uptake and release rates of epiphytic algae [*51,56*]. As it has been demonstrated that phytoplankton productivity in acidic lakes can be stimulated by phosphorus addition [*8–11*], it is conceivable that structural changes in the littoral zone during acidification, including the expansion of epiphytic algae and the often observed demise of submergent angiosperms, can alter phosphorus availability in the pelagic zone and thus play a part in the observed changes in phytoplankton structure and abundance with acidification.

Although poorly understood, the species composition of littoral macrophyte communities may affect water chemistry. Replacement of submergent angiosperms in temperate acidic lakes by the bryophyte *Sphagnum* may promote further acidification of a lake through the latter's ion exchange capacity where calcium, magnesium, iron, sodium, and potassium are removed from the water column in exchange for hydrogen ions [*39,57*]. From an ongoing investigation of 14 macrophyte species collected from 20 acidic Florida lakes (pH 4.4 to 6.4), our preliminary results suggest that trace elements, including copper, cadmium, lead, and zinc, are generally concentrated by plants with acidification, but the source for these elements (sediments or overlying water column) is highly species specific.

Influence of Littoral Vegetation on Zooplankton and Benthic Invertebrates

While the structure and extent of littoral and benthic autotrophic communities are likely to greatly influence total ecosystem productivity, their indirect control on observed zooplankton, benthic invertebrate, and fish responses to acidification are poorly known. As discussed earlier, both the biomass and diversity of pelagic zooplankton decline with acidification, but it has not been possible to separate the direct effects of pH from those related to food availability. Fryer [*58*], based on collections of planktonic and littoral fauna from 70 acidic water bodies in the United Kingdom, concluded that crustacean zooplankton diversity decreases with increasing acidity. While the work of Fryer is informative, no attempt was made to separate the contribution of littoral versus planktonic communities to the observed relationship.

With the exception of changes in the diversity of chydorid cladoceran assemblages, the effect of increasing acidity on the structure of littoral zooplankton communities is largely unknown. Littoral assemblages of chydorid cladocerans appear to be less impacted by acidity than crustacean zooplankton in the pelagic zone. Both Whiteside [59] and Crisman [60] reported highest chydorid diversity in soft-water oligotrophic lakes. Although Brakke [61] suggested an opposite trend, his data showed that chydorid diversity was reduced in only 5 of 12 acidic (pH 4.4 to 5.2) Norwegian lakes relative to less acidic (pH 5.7 to 6.7) lakes. While no attempt has been made to relate observed chydorid diversity along acidity gradients to the structure of littoral and benthic autotrophic communities, Whiteside et al. [62] have shown that within a single lake chydorid abundance and diversity are greater for macrophyte than for bare mud habitats. Detailed examination of pelagic and littoral zooplankton simultaneously from a series of lakes would help to elucidate the response of total lake zooplankton to acidification and may shed some light on the importance of direct pH effects, food availability, and changes in predation pressure on observed relationships.

Detailed literature reviews on the response of benthic invertebrate communities to acidification recently have been published [36,45,63]. From these reviews, it is apparent that, while earlier investigations stressed direct pH effects as controlling observed changes in benthos community structure with increasing acidity, recent studies have placed increasing importance on the indirect effects of pH on food availability and predation pressure. Although the latter are strongly influenced by habitat availability, only Wiederholm and Eriksson [35], and Singer [45] have recognized that alterations in the structure of littoral and benthic autotrophic communities may be important in observed benthos responses to acidity. This relationship is currently in need of quantification.

Influence of Littoral Vegetation on Fish Communities

Earlier it was stated that a major reduction in the species richness and biomass of fish communities has been noted with increasing acidity in both temperate and subtropical Florida lakes. A progressive loss of individual species with increasing acidity has been used to account for the reduction in species richness in temperate lakes, while such an interpretation cannot be applied to the similar trend observed in comparable Florida lakes because no species are lost with acidity. The elimination of fish species in acidic temperate lakes has been attributed to both acute and chronic direct pH effects and pH-related changes in the availability of sodium, calcium, aluminum, and heavy metals [2,24–25]. Comparably acidic Florida lakes have similar sodium and calcium concentrations, much lower aluminum (<100 μg/L) and heavy metal levels, and lack major episodic reductions in pH, such as documented for temperate lakes during spring snow melt. Thus, while chemistry is likely the ultimate factor controlling fish responses to increasing acidity, reductions in both biomass and species richness of Florida fish communities in spite of the lack of a chemically-related elimination of individual species suggest that other nonchemical factors also play a role in the observed relationships. Foremost among the latter factors is likely to be the structure of littoral and benthic autotrophic communities.

DiCostanzo [64] noted that the success of bluegill year classes depended on the presence of a vegetated littoral zone, and Guillory et al. [65], through a comparison of fish communities from vegetated and nonvegetated (from residential development) littoral areas within individual Florida lakes, found that fish diversity and biomass were positively related to the extent of macrophytes. Later work by Crowder and Cooper [66] showed that the best bluegill growth was associated with moderate macrophyte density. They concluded that macrophytes were needed as a refuge from fish predators, but that the reduced growth of bluegills in extremely dense macrophyte beds was related to the difficulty that the fish had in catching prey.

While it is generally agreed that benthic invertebrate abundance and diversity are greatest in vegetated littoral zones [66], the importance of this potential food resource for freshwater fish is often not clear. Some investigators [66–67] have shown that young centrarchids, especially bluegill and bass, can exert strong seasonal predation pressure on micro- and macroinvertebrate communities in the vegetated littoral zone, causing major species shifts and increased dominance of small-bodied taxa. Hall and Werner [68] suggested that mature centrarchids overwinter in vegetated areas, in part because of greater invertebrate abundance there. On the other hand, Fairchild [69] felt that seasonal predation by bass fry affected the cladoceran *Sida* but had no noticeable impact on either chironomid or chydorid cladoceran communities. This general view was supported by Werner et al. [70], who noted that, although benthic invertebrates are most diverse and abundant in macrophytes (in this case, *Typha*), bluegills generally feed elsewhere.

The importance of macrophytes as a predation refugium for young fish is clear. Bluegills <100 mm stay in vegetated areas to avoid bass predation [66,68,71–72]. Fairchild [69] noted the same behavior for bass fry, and Mittelbach [71] suggested that most fish species act similarly. Both Mittelbach [71] and Werner et al. [72] stressed that in spite of the potential for maximizing their energy gain by foraging in the pelagic zone, small bluegill are behaviorally restricted to the vegetated littoral zone in order to avoid intense bass predation in open water.

Thus, alteration of either the aerial extent or structure of the vegetated littoral zone can have a profound impact on the species composition, diversity, biomass, growth rate, and age structure of fish communities. While chemistry is likely the ultimate factor controlling fish responses to increasing acidity, a similar reduction in both species richness and biomass in acidic Florida lakes without the progressive elimination of individual species observed in comparably acidic temperate lakes suggests that other covarying factors also may contribute to observed relationships. Of these, structural changes in the littoral zones on size-specific predation and interspecific competition for breeding sites and food resources are likely to play a major role in the observed simplification of fish communities.

A vast majority of investigations examining changes in aquatic biota relative to increasing acidity have concentrated on pelagic components. The possible influence of littoral-pelagic interactions on observed relationships has largely been ignored, and thus interpretation of pelagic biotic data unfortunately is often taken out of context of the whole ecosystem. While littoral and benthic communities are major contributors to the autotrophic production of small lakes in general, it is likely that their share assumes greater importance in acidic lakes where phytoplankton are often severely nutrient limited. In turn, the structure of the littoral zone can have an important indirect influence on heterotrophic communities principally through habitat and food quality and quantity. Future investigations of biota in acidic soft-water lakes should concentrate on quantifying such littoral-pelagic linkages.

References

[1] Almer, B., Dickson, W., Ekstrom, C., Hornstrom, E., and Miller, U., *Ambio*, Vol. 3, 1974, pp. 30–36.
[2] Leivestad, H., Hendrey, G., Muniz, I. P., and Snekvik, E. in *Impact of Acid Precipitation on Forest and Freshwater Ecosystems in Norway*, FR 6/76, SNSF project, As, Norway, 1976, pp. 87–111.
[3] Yan, N. D. and Stokes, P. M., *Environmental Conservation*, Vol. 5, 1978, pp. 93–100.
[4] Kwiatkowski, R. E. and Roff, J. C., *Canadian Journal of Botany*, Vol. 54, 1976, pp. 2546–2561.
[5] Almer, B., Dickson, W., Ekstrom, C., and Hornstrom, E. in *Sulfur in the Environment. Part II. Ecological Impacts*, J. O. Nriagu, Ed., Wiley, New York, 1978, pp. 271–311.
[6] "Acidification in the Canadian Aquatic Environment: Scientific Criteria for Assessing the Effects of Acid Deposition on Aquatic Ecosystems," NRCC/CNRC, National Research Council Canada, Ottawa, 1981.
[7] Hendrey, G. R., Baalsrud, K., Traaen, T. S., Laake, M., and Raddum, G., *Ambio*, Vol. 5, 1976, pp. 224–227.
[8] Dillon, P. J., Yan, N. D., Scheider, W. A., and Conroy, N., *Archiv fur Hydrobiologie Beihefte Ergebnis Limnologie*, Vol. 13, 1979, pp. 317–336.
[9] Fee, E. J., *Canadian Journal of Fisheries and Aquatic Sciences*, Vol. 37, 1980, pp. 513–522.
[10] Wilcox, G. and DeCosta, J., *Archiv fur Hydrobiologie*, Vol. 94, 1982, pp. 393–424.
[11] DeCosta, J., Janicki, A., Shellito, G., and Wilcox, G., *Oikos*, Vol. 40, 1983, pp. 283–294.
[12] Yan, N. D., *Water Air Soil Pollution*, Vol. 11, 1979, pp. 43–55.
[13] Roff, J. R. and Kwiatkowski, R. G., *Canadian Journal of Zoology*, Vol. 55, 1977, pp. 899–911.
[14] Confer, J. L., Kaaret, T., and Likens, G. E., *Canadian Journal of Fisheries and Aquatic Sciences*, Vol. 40, 1983, pp. 36–42.
[15] Schindler, D. W. and Noven, B., *Journal of the Fisheries Research Board of Canada*, Vol. 28, 1971, pp. 245–256.
[16] Beaver, J. R. and Crisman, T. L., *Verhandlungen Internationale Vereinigung Limnologie*, Vol. 21, 1981, pp. 353–358.
[17] Beaver, J. R. and Crisman, T. L., *Liminology and Oceanography*, Vol. 27, 1982, pp. 246–253.
[18] Brezonik, P. L., Crisman, T. L., and Schulze, R. L., *Canadian Journal of Fisheries and Aquatic Sciences*, Vol. 41, 1984, pp. 46–56.

[19] Wright, R., Dale, T., Gjessing, E., Hendrey, G., Henricksen, A., Johannessen, M., and Muniz, I., *Water Air Soil Pollution,* Vol. 6, 1976, pp. 483–499.
[20] Almer, B., *Forsurningens Inverkan Pa Fiskbestand I Vastkutsjoar,* Information fran Sotuattens-Laboraotiret, Drottningholm, No. 12, 1972.
[21] Harvey, H. H., *Verhandlungen Internationale Vereinigung Limnologie,* Vol. 19, 1975, pp. 2406–2417.
[22] Schofield, C. L., United States Forest Service, General Technical Report NE-23, 1976, pp. 477–478.
[23] Keller, A. E., "Fish Communities in Florida Lakes: Relationship To Physico-Chemical Parameters," Master of Science thesis, University of Florida, Gainesville, 1984.
[24] Baker, J. P. and Schofield, C. L., *Water Air Soil Pollution,* Vol. 18, 1982, pp. 289–309.
[25] Beamish, R. J., United States Forest Service, General Technical Report NE-23, 1976, pp. 479–498.
[26] Traaen, T. S., Technical Note TN 41/78, SNSF project, Norway, 1978.
[27] Traaen, T. S. in *Proceedings of International Conference on the Ecological Impact of Acid Precipitation,* D. Drablos and A. Tollan, Eds., SNSF project, Norway, 1980, pp. 340–341.
[28] Muller, P., *Canadian Journal of Fisheries and Aquatic Sciences,* Vol. 37, 1980, pp. 355–363.
[29] Boylen, C. W., Shick, M. D., Roberts, D. A., and Singer, R., 1983, *Applied Environmental Microbiology,* Vol. 45, pp. 1538–1544.
[30] Scheider, W., Adamski, J., and Paylor, M., *Ontario Ministry of the Environment Report,* Rexdale, Ontario, 1975.
[31] Scheider, W. and Dillon, P. J., *Water Pollution Research Control,* Vol. 11, 1976, pp. 93–111.
[32] Rao, S. S. and Dutka, B. J., *Hydrobiologia,* Vol. 98, 1983, pp. 153–157.
[33] Crisman, T. L., Scheuerman, P., Bienert, R. W., Beaver, J. R., and Bays, J. S., *Verhandlungen Internationale Vereinigung Limnologie,* Vol. 22, 1984, pp. 620–626.
[34] Okland, J. and Okland, K., in *Proceedings International Conference on the Ecological Impact of Acid Precipitation,* D. Drablos and A. Tollan, Eds., SNSF project, As, Norway, 1980, pp. 326–327.
[35] Wiederholm, T. and Eriksson, L., *Oikos,* Vol. 29, 1977, pp. 261–267.
[36] Haines, T. A., *Transactions of the American Fisheries Society,* Vol. 110, 1981, pp. 669–707.
[37] Collins, N. C., Zimmerman, A. P., and Knoechel, R. in *Effects of Acidic Precipitation on Benthos,* R. Singer, Ed., North American Benthological Society, Springfield, IL, 1981, pp. 35–48.
[38] Grahn, O., *Water Air Soil Pollution,* Vol. 7, 1977, pp. 295–306.
[39] Hendrey, G. R. and Vertucci, F. A. in *Proceedings of the International Conference on the Ecological Impact of Acid Precipitation,* D. Drablos and A. Tollan, Eds., SNSF project, As, Norway, 1980, pp. 314–315.
[40] Nilssen, P., *Internationale Revue Gesamten Hydrobiologie,* Vol. 65, 1980, pp. 177–207.
[41] Roelofs, J. G. M., *Aquatic Botany,* Vol. 17, 1983, pp. 139–155.
[42] Lazarek, S., Research Report, DNR-403-2328-78, "Fiskeristyrelsen," Lund, Sweden, 1979.
[43] Lazarek, S., *Naturwissenschaften,* Vol. 67, 1980, p. 97.
[44] Schindler, D. W. and Turner, M. A., *Water Air Soil Pollution,* Vol. 18, 1982, pp. 259–271.
[45] Singer, R. in *Acid Precipitation: Effects on Ecological Systems,* Chapter 15, F. M. D'Itri, Ed., Ann Arbor Science, MI, 1982, pp. 329–363.
[46] Stokes, P. M. in *Effects of Acidic Precipitation on Benthos,* R. Singer, Ed., North American Benthological Society, Springfield, IL, 1981, pp. 119–138.
[47] Grahn, O., Hultberg, H., and Landner, I., *Ambio,* Vol. 3, 1974, pp. 93–94.
[48] Andersson, G., Fleischer, S., and Graneli, W., *Verhandlungen Internationale Vereinigung Limnologie,* Vol. 20, 1978, pp. 802–807.
[49] Schindler, D. W., Wagemann, R., Cook, R. B., Ruszczynski, T., and Prokopowich, J., *Canadian Journal of Fisheries and Aquatic Sciences,* Vol. 37, 1980, pp. 342–354.
[50] Gahnstrom, G., Andersson, G., and Fleischer, S. in *Proceedings of the International Conference on the Ecological Impact of Acid Precipitation,* D. Drablos and A. Tollan, Eds., SNSF project, As, Norway, 1980, pp. 306–307.
[51] Wetzel, R. G., *Limnology,* 2nd ed., Saunders, Philadelphia, 1983.
[52] Canfield, D. E., Jr., Shireman, J. V., Colle, D. E., Haller, W. T., Watkins, C. E., II, and Maceina, M. J., *Canadian Journal of Fisheries and Aquatic Sciences,* Vol. 41, 1984, pp. 497–501.

[53] Schindler, D. W. in *Proceedings of the International Conference on the Ecological Impact of Acid Precipitation*, D. Drablos and A. Tollan, Eds., SNSF project, As, Norway, 1980, pp. 370–374.
[54] Barko, J. W. and Smart, R. M., *Freshwater Biology*, Vol. 10, 1980, pp. 229–238.
[55] Carignan, R. and Kalff, J., *Science*, Vol. 207, 1980, pp. 987–989.
[56] Confer, J. L., *Ecological Monographs*, Vol. 42, 1972, pp. 1–23.
[57] Hultberg, H. and Grahn, O., *International Association of Great Lakes Research*, Supplement Vol. 2, 1975, pp. 208–217.
[58] Fryer, G., *Freshwater Biology*, Vol. 10, 1980, pp. 41–45.
[59] Whiteside, M. C., *Ecological Monographs*, Vol. 40, 1970, pp. 79–118.
[60] Crisman, T. L. in *Evolution and Ecology of Zooplankton Communities*, W. C. Kerfoot, Ed., University Press of New England, 1980, pp. 657–667.
[61] Brakke, D. F. in *Proceedings of the International Conference on the Ecological Impact of Acid Precipitation*, D. Drablos and A. Tollan, Eds., SNSF project, As, Norway, 1980, pp. 272–273.
[62] Whiteside, M. C., Williams, J. B., and White, C. P., *Ecology*, Vol. 59, 1978, pp. 1177–1188.
[63] Hendrey, G. R., Yan, N. D., and Baumgartner, K. J. in *Restoration of Lakes and Inland Waters*, US EPA 440/5-81-010, 1980, Environmental Protection Agency, Washington, DC, pp. 457–466.
[64] DiCostanzo, C. J., *Iowa State College Journal of Science*, Vol. 32, 1957, pp. 19–34.
[65] Guillory, V., Jones, M. D., and Rebel, M., *Florida Scientist*, Vol. 42, 1979, pp. 113–122.
[66] Crowder, L. B. and Cooper, W. E., *Ecology*, Vol. 63, 1982, pp. 1802–1813.
[67] Phoenix, D., "Temporal Dynamics of a Natural Multipredator-Multiprey System," Ph.D. thesis, University of Pennsylvania, Philadelphia, 1976.
[68] Hall, D. J. and Werner, E. E., *Transactions of the American Fisheries Society*, Vol. 106, 1977, pp. 545–555.
[69] Fairchild, G. W., *Hydrobiologia*, Vol. 96, 1982, pp. 169–176.
[70] Werner, E. E., Mittelbach, G. G., Hall, D. J., and Gilliam, J. F., *Ecology*, Vol. 64, 1983, pp. 1525–1539.
[71] Mittelbach, G. G., *Ecology*, Vol. 62, 1981, pp. 1370–1386.
[72] Werner, E. E., Gilliam, J. F., Hall, D. J., and Mittelbach, G. G., *Ecology*, Vol. 64, 1983, pp. 1540–1548.

Lise Parent,[1] Martine Allard,[2] Dolors Planas,[1] and Guy Moreau[2]

The Effects of Short-Term and Continuous Experimental Acidification on Biomass and Productivity of Running Water Periphytic Algae

REFERENCE: Parent, L., Allard, M., Planas, D., and Moreau, G., "**The Effects of Short-Term and Continuous Experimental Acidification on Biomass and Productivity of Running Water Periphytic Algae,**" *Impact of Acid Rain and Deposition on Aquatic Biological Systems, ASTM STP 928*, B. G. Isom, S. D. Dennis, and J. M. Bates, Eds., American Society for Testing and Materials, Philadelphia, 1986, pp. 28–41.

ABSTRACT: The effect of short-term and continuous acidification on the primary production and biomass of periphyton was studied in seminatural conditions using an experimental device simulating streams ecosystems. After allowing periphytic algae to colonize unglazed ceramic tiles for 4 to 6 weeks under natural conditions, the pH of the water was lowered with sulfuric acid (H_2SO_4) from 6.5 to 6.7 to 4 to 4.6 for short periods (12 to 72 h) or long periods (continuous acidification, 84 days). During short-term acidification, in all the experiments but one, there was a significant increase (up to 2.6 times as compared to the control) of the primary production and specific activity in the first 12 h, followed by a significant decrease and slow recovery from 48 h until the end of the experiment. If acidification is continuous, primary production and specific activity are significantly higher (between 10 to 15 times that of the control) in acidified channels, either with or without aluminum, by Day 33 and until the end. Biomass stayed constant or decreased in short-term acidification and increased (from 10 to 15 times) in continuous acidification.

KEY WORDS: experimental acidification, periphyton, running water, productivity, biomass

Acid precipitation comprises hydrogen ions, nutrients, and other pollutants, which are intermittently released into ecosystems with each precipitation event. Each rain produces two possible impacts: an immediate one (pH drop, nutrient loading) and a delayed one through seepage and percolation of water through soil, resulting in heavy metal lixiviation.

Large rivers and lakes having a relatively good buffering capacity are only slightly affected by sporadic precipitation except, of course, during spring snow

---

[1] Université du Quebec a Montreal, Departement des Sciences Biologiques, Montreal, P.Q., Canada H3C–3P8.
[2] Université Laval, Departement de Biologie, Quebec, P.Q., Canada G1K–7P4.

melt. Smaller streams and ponds, however, because of their poorer buffering capacity, are much more sensitive to this type of disturbance. The continuous acid loading, or poor buffering capacities, can result in pH decreases which can lead an ecosystem into an irreversible acidification process comparable to the titration of a base with an acid [1]. Consequently, aquatic communities are submitted to two different phenomena: (1) a temporary stress and (2) a continuous slow evolution of the milieu. The temporary stress also could occur in ecosystems under continuous acidification after a rain or snow melt.

The effect of short-term acidification occurs mainly in running waters where the bottom communities (benthos and periphyton) are dominant.

Existing literature deals with the effects of long-term acidification on aquatic communities, but no study takes into account the two types of stress: temporary and continuous.

For this reason, the principal objectives of our study are:

1. To establish the impact caused by each kind of stress.
2. To verify if the impact caused by short-term acidification is reversible.
3. To find out if the modifications caused by short-term acidification could explain the effects of continuous acidification.

Our secondary objectives are:

1. To compare the action of acid alone to the action of acid plus aluminum.
2. To test if predation could explain the higher biomass of periphytic algae.

To reach these objectives, we conducted a study in a controlled environment in which the effects of environmental modifications to different trophics levels were investigated.

Existing literature deals with the effects of long-term acidification on aquatic communities, mainly on the upper trophic levels, and a few are concerned with the primary producers (see, for example, the Haines review [2]).

Ecological effects of acid precipitation on primary producers were reviewed [3-5]. The striking point in these reviews is the general lack of agreement in the response of primary production or biomass algae.

Phytoplankton biomass and primary production are generally lower in acidic lakes [6-8], and, inversely, in most cases periphyton biomass appears to be higher [9]; dense mats of filamentous algae in lakes [10,11] and in rivers [5,12] were reported. Hall et al. [13] also noticed a visually apparent increase in periphyton in a section of a brook experimentally acidified. Although contradictory results are noted in several studies, Schindler et al. [14] found that, after two years of acidification of Lake 223, phytoplanktonic photosynthesis and biomass were not affected. The same author [15] reported a significant increase in biomass after 5 years of acidification of the same lake. Experiments on periphyton in enclosures [9] showed that periphyton biomass appeared to be higher in the acidified enclosure and that primary production was not enhanced. Comparisons of acidic to neutral Pennsylvania streams failed to reveal any difference in primary production or chlorophyll in the periphyton [16]. Finally, production and biomass of periphytic algae were higher in experimentally acidified channels, although the specific activity was lower at pH 4.0 than at the natural pH [12].

In our opinion, the reasons for these contradictory results are:

1. The diversity of community structure in different ecosystems historically exposed to different pollutants.
2. The difficulty to know how long the ecosystems under study had been submitted to natural acidification.
3. The lack of data on the effects of intermittent pH drops on the structure and metabolism of communities.
4. The difficulty of separating, in natural ecosystems, the effects of hydrogen ions from that of the metals, whose solubility and chemical speciation vary with pH and other chemical characteristics of water.

To avoid, at least partially, the inconvenients just mentioned, we have utilized an experimental device that allows us to work with communities resembling natural ones while being able to manipulate some of the parameters (pH, aluminum concentration, flow, sediment), providing us with a real control. At the same time we can study, in similar conditions, the effect of short-term and continuous acidification on periphytic communities.

Material and Methods

The experimental wooden troughs used for short-term and continuous acidification are described elsewhere [17]. They were set up in Forêt Montmorency, Laval University Research Station, located 100 km north of Quebec City (47°19′ north latitude, 71°07′ west longitude) in a boreal forest dominated by balsam fir (*Abies balsamea*) in association with white birch (*Betula papyrifera*). This region is characterized by abundant total precipitation (the annual mean for the last 15 years was 1500 mm [18]); the mean pH of the rain was 4.3 and the sulfate (SO_4) and nitrogen oxide (NO_x) concentrations were 41.1 and 26.8 μeq L^{-1}, respectively [19]. Of the two trough sets [17], three troughs were used for short-term acidification and three for long-term acidification experiments. In both experiments, the artificial substratum used for periphyton colonization was unglazed ceramic tiles having a surface of 10 cm^2. In order to have all tiles with the same orientation, they were placed over a plastic support with an inclination of 45°. The bottom of the channel was covered with pebbles. Before treatment, the tiles were left to colonize by the organisms drifting in the creek [the faunistic composition of communities is described in Ref 20 and in Footnote 5 (see page 38)] from 4 to 6 weeks (previous experimentation showed that 3 weeks was enough to obtain a periphytic biomass representative of the creeks in the region). For the short-term acidification, after the initial colonization, the communities were left in place from one treatment to another.

In both experiments, acidification was done with sulfuric acid (H_2SO_4) 3.5 N to decrease the pH to 4 to 4.6. In the continuous acidification experiment, in one of the channels, in addition to H_2SO_4, 400 µg L^{-1} of aluminum sulfate were added. In both setups, one channel was kept as a control. For short-term acidification the acid was added for a period of 12 to 72 h several times during the summer.

In each short-term treatment, pH, conductivity, temperature, light, and oxygen were measured every 4 h. Nutrients, cations, anions, and metals were sampled daily, and the analyses were done in duplicates. In the case of continuous acidification, pH, conductivity, and temperature were measured at the same hour every day. Samples for nutrients, cations, anions, and metals were taken at the same time as biological samples. The analytical procedures for these parameters are listed in Table 1.

For short-term acidification, a total of nine experiments were performed between June and September in 1982 and 1983. Biomass (chlorophyll-a) and primary production were analyzed the day before the treatment (-24 h), at the beginning of acidification (0 h), and 4, 24, 48, and 72 h after the start of acidification and

TABLE 1—*Analytical methods used in both experiences.*

Analysis	Method
Color	colorimetric [44]
Conductivity	conductivity meter Horizon 1484-10
pH	pH meter Metrohm AG CH 9100
Alkalinity	unfiltered water: titrimetry [45] or conductivity meter titration [44]
Total inorganic carbon (TIC)	unfiltered water: titrimetry [45] or carbon infrared detector [44]
Total organic carbon (TOC)	unfiltered water: oxydation with cobalt oxide or infrared detector [44]
Ammonia nitrogen (N − NH_4)	unfiltered water: indophenol reaction [44]
Nitrate + Nitrite (N − NO_3 + NO_2)	filtered water cadmium reduction of nitrates [44][3]
Soluble reactive phosphorus (P − PO_4)	filtered water: acid ascorbic reduction [46][3]
Total phosphorus (P − TP)	unfiltered water: potassium persulfate oxydation, acid ascorbic reduction [46][3]
Soluble reactive Silicon (Si − SiO_2)	unfiltered water: blue molybdate [44]
Sulfate (SO_4^{2-})	filtered water methylthymol blue method [44] or ion chromatograph Dionex 2010
Chloride (Cl^{-1})	filtered water: thiocyanate [44] or ion chromatograph Dionex 2010
Ca, Na, K, Mg, Zn, Cu, Fe	atomic absorption [44]
Aluminum (Th, Al, TDAl)	plasma emission [44]

[3] Technicon Auto Analyser II, methodologie, Technicon Industrial System, Tarrytown, NY, No. 181-72W, 19-69W, 158-71W, 94-70W, 105-71W, 118-71W, 99-70W, 1973.

also 24-h postacidification (+24 h). In the continuous acidification experiment (from June to October), samples were taken the day before and the day after acidification, and a month (Day 33), two months (Day 55), and three months (Day 84) thereafter.

Primary production was measured using the ^{14}C method. Each tile was placed in one 250-mL Pyrex bottle full of water; 0.5 mL of an 8-μCi ^{14}C-bicarbonate was added (the addition of this solution increased the pH of the acidified water from 0.05 to 0.1 unit only). For each channel, three clear and two black bottles were placed in a shaker and incubated in the creek water. The site chosen for incubation was receiving similar irradiance as the channels. At the end of the incubation period, samples were put in a dark cooler, and the algae were immediately scraped with a brush into the incubation bottles. This suspension was mixed thoroughly, and a 50-mL subsample was filtered through a 0.45-μm Millipore filter. Two mL of hydrochloric acid (HCl) 0.001 N were added at the end of filtration in order to dissolve any possible ^{14}C-carbonate precipitation [21]. The filters were put in a scintillation cocktail [22] and counted with a Beckman scintillation counter. To know the total available ^{12}C, carbon dioxide (CO_2) was measured titrimetrically.

From three to five tiles were used for the chlorophyll-a extraction with 99% methanol [23,24]. Specific activity was calculated as the ratio of mean primary production over mean chlorophyll-a, expressed in mgC mgChla^{-1}h^{-1}.

The experiment for predation assessment was done in continuous acidification channels at the end of September. Larvae of the families Leptophlebiidae and Ephemerellidae were collected in the creek and put for a 1-day acclimation period in the test water. Then 14 larvae were put in a plastic tube (15 cm long, 3 cm diameter) with small ceramic tiles (2 by 2 cm), colonized since the beginning of the experiment. Tube extremities were closed with Nytex (250 μm). Tubes with and without mayflies were put in the channel for 14 days. Twelve tiles in tubes with mayflies and 6 in tubes without mayflies were randomly sampled in each channel, and chlorophyll-a was measured with the method just described.

Mann and Whitney and Kruskal-Wallis tests were used to compare primary production and chlorophyll-a among treatments at each sampling period.

Results

Chemical parameters

The waters flowing through our experimental devices have a pH close to neutrality, and their chemical characteristics correspond to those of oligotrophic ecosystems in the Canadian Shield: low in conductivity, poor in nutrients, and relatively rich in organic matter (Tables 2 and 3).

The addition of acid modified mainly the parameters directly related to this accretion: increase in SO_4^{2-} and conductivity and a decrease in pH and alkalinity. No variation was observed in the nutrient concentrations and in major cations: calcium, manganese, and sodium. Potassium increased during the treatment, but

TABLE 2—*Main chemical parameters in the short-term acidification experiment. Range of concentrations during treatment. Symbols are the same as in Table 1.*

	Control	pH 4.0
pH	6.1 to 7.1	4.1 to 4.6
Conductivity ($\mu S\ cm^{-1}$)	25.0 to 39.0	55.0 to 102.0
TIC (mg L^{-1})	0.9 to 1.8	0.9 to 1.3
TOC (mg L^{-1})	4.6 to 8.6	3.4 to 7.1
N – NO_2 + NO_3 ($\mu g\ L^{-1}$)	38.5 to 207.0	67.3 to 230.0
P – PO_4 ($\mu g\ L^{-1}$)	<0.9 to 6.8	<0.9 to 5.9
P – PT ($\mu g\ L^{-1}$)	6.8 to 40.9	5.9 to 17.7
SO_4^{2-} (mg L^{-1})	1.4 to 4.3	6.3 to 11.2
Cl^{1-} (mg L^{-1})	0.2 to 1.8	0.3 to 1.5
Ca^{2+} (mg L^{-1})	0.4 to 0.9	0.4 to 0.9
Mg^{2+} (mg L^{-1})	0.4 to 0.7	0.4 to 0.7
Na^+ (mg L^{-1})	1.0 to 1.9	1.0 to 1.9
K^+ (mg L^{-1})	0.001 to 0.09	0.08 to 0.4
Fe ($\mu g\ L^{-1}$)	<10.0 to 155.0	<10.0 to 154.0
Zn ($\mu g\ L^{-1}$)	<2.0 to 140	<2.0 to 110
TDAl ($\mu g\ L^{-1}$)	85.0 to 125.0	100.0 to 185.5

TABLE 3—*Main chemical parameters in the continuous acidification experiment. Range of concentrations during treatment. Symbols as in Table 1.*

	Control	pH 4.0	pH 4.0 + Al
pH	6.3 to 6.8	4.0	4.0
Alkalinity (mg L^{-1} $CaCO_3$)	3.1 to 5.3	<0.1	<0.1
Conductivity ($\mu S\ cm^{-1}$)	19.0 to 21.0	58.0 to 63.0	62.0 to 72.0
Color (Hazen unit)	13.0 to 17.0	10.0 to 13.0	11.0 to 14.0
TIC (mg L^{-1})	1.0 to 1.5	<0.5 to 0.5	<0.5 to 0.5
TOC (mg L^{-1})	3.0 to 7.0	4.0 to 7.5	3.5 to 9.5
N – NO_2 + NO_3 ($\mu g\ L^{-1}$)	10.0 to 30.0	10.0 to 20.0	10.0 to 30.0
N – NH_4 ($\mu g\ L^{-1}$)	<20.0 to 20.0	<20.0 to 20.0	<20.0 to 20.0
P – PO_4 ($\mu g\ L^{-1}$)	<6.5	<6.5	<6.5
P – TP ($\mu g\ L^{-1}$)	<6.5 to 9.8	<6.5 to 6.5	<6.5 to 9.8
Si – SiO_2 ($\mu g\ L^{-1}$)	8.6 to 12.4	8.6 to 12.4	8.6 to 12.4
SO_4^{2-} (mg L^{-1})	3.0 to 3.5	11.0 to 13.0	12.0 to 14.5
Cl^{1-} (mg L^{-1})	0.2 to 0.4	0.2 to 0.6	0.2 to 0.6
Ca^{2+} (mg L^{-1})	2.1 to 2.6	2.1 to 2.7	2.1 to 2.6
Mg^{2+} (mg L^{-1})	0.4 to 0.6	0.5 to 0.6	0.5 to 0.6
Na^+ (mg L^{-1})	0.9 to 1.6	1.0 to 1.6	0.8 to 1.2
K^+ (mg L^{-1})	0.2 to 0.5	0.2 to 0.5	0.2 to 0.4
Fe ($\mu g\ L^{-1}$)	50.0 to 150.0	50.0 to 140.0	50.0 to 210.0
Cu ($\mu g\ L^{-1}$)	<5.0 to 9.0	<5.0 to 13.0	<5.0 to 9.0
Zn ($\mu g\ L^{-1}$)	<10.0 to 60.0	<10.0 to 60.0	<10.0 to 50.0
TAl ($\mu g\ L^{-1}$)	50.0 to 180.0	60.0 to 220.0	230.0 to 520.0
TDAl ($\mu g\ L^{-1}$)	50.0 to 130.0	60.0 to 140.0	230.0 to 520.0

34 IMPACT OF ACID RAIN

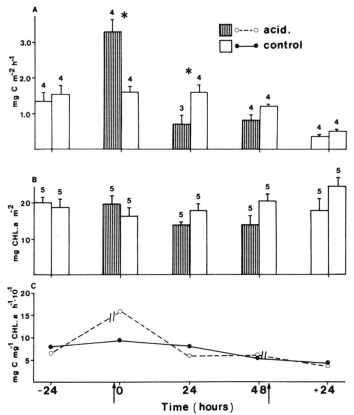

FIG. 1—*Short-term acidification. General case ($P\text{-}PO_4 > 2.8$ µg L^{-1}): (A) Primary production; (B) Chlorophyll-a; (C) Specific activity. Starting and ending of the acidification period are indicated by arrows. For A and B, X ± 1 standard error (SE), the number of replicates is indicated above the mean. Significant differences between acid and control are indicated *($p < 0.05$).*

only in the short-term acidification experiment. Of all metals measured (iron, manganese, zinc, copper, cadmium, nickel, aluminum, chromium, and lead), only aluminum, zinc, and iron were detectable in both type of experiments, and copper only in continuous acidification. No differences in the concentrations of aluminum, zinc, and copper were noticed between the treated and the control water. In the channel where aluminum was added, the maximum value of iron was slightly higher than in the other two channels. Color decreased in the continuously acidified channel as total inorganic carbon (TIC). In the short-term experiment color was not measured.

Biological Parameters, Short-Term Acidification

The pattern of evolution of biomass and primary production has been the same in eight of the nine experiments conducted during 1982 and 1983. This pattern is presented in Fig. 1.

In the first 4 h after acidification started, biomass was not significantly different in control and acidified channels (Fig. 1B). Periphytic production, however, was significantly higher (Fig. 1A) in the presence of acid; specific activity is therefore higher (Fig. 1C). Twenty-four hours after the start of the acidification, biomass did not vary but production became significantly lower ($P < 0.05$) in the presence of H_2SO_4. At the 48-h point, no differences were observed with regard to biomass or production. This situation continued until the 24-h postacidification point.

The sole exception to this pattern was seen in the experiment where soluble reactive phosphorus was nondetectable (July 1983). In this situation, the increase in production, immediately following the start of acidification, was not observed, and production levels stayed identical to the control at all sampling periods: 4, 24, 48, and 72 h after the start of the acidification and also the 24-h postacidification point (Fig. 2A). Biomass, however, was always significantly lower ($P < 0.05$ or $P < 0.01$) in acidified channels from the 4-h point until the end of the experiment (Fig. 2B). Specific activity of algae in acidified channels was higher from 48 h after the start of acidification until the end of the experiment (Fig. 2E).

Biological Parameters, Continuous Acidification

One day after acidification, primary production was significantly lower in the treated channels than in the control (Fig. 3A). Meanwhile, no significant difference was observed for biomass (Fig. 3B).

On Days 33, 55, and 84 a clear pattern appears: primary production and biomass were significantly higher ($P < 0.05$) in acidified channels. No significant difference between acidified channels, with or without aluminum, was noticed.

Specific activity varies through time and among channels. For the control, it increases one day after acidification, and thereafter it decreases steadily. For the acidified channels, specific activity is lowered at the beginning, and from Day 1 (for acid plus aluminum) or Day 33 (for acid) it increases until the end of experiment. At that time, the specific activity was nearly ten times higher in both acidified channels compared to the control (Fig. 3C).

Predation Assessment

In all three conditions, chlorophyll-a levels were slightly higher on the tiles incubated without mayflies; however, the differences were not statistically significant ($P > 0.05$) (Fig. 4).

Discussion

In our experimental conditions, a negative effect of acidification on biomass and periphytic production appeared only at the 24-h acidification point. This response is short-lived since, in eight of nine experiments, differences have disappeared between experimental conditions and controls, 24 h after the end of acidification, even with algae that had undergone previous acidification events.

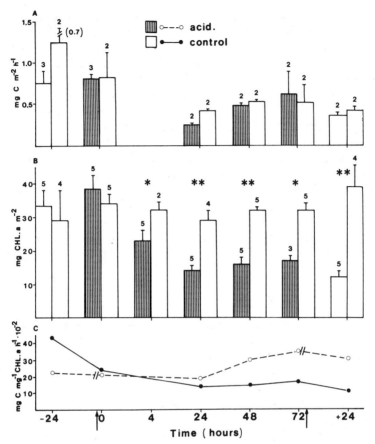

FIG. 2—*Short-term acidification. When $P-PO_4$ is below detectable limit ($<0.9 \mu g L^{-1}$): (A) Primary production; (B) Chlorophyll-a; (C) Specific activity. Starting and ending of the acidification period are indicated by arrows. For A and B, $X \pm 1$ standard error (SE), the number of replicates is indicated above the mean. Significant differences beween acid and control are indicated *($p < 0.05$), **($p < 0.01$).*

Inversely, production is stimulated for a very brief period immediately following the start of acidification and after a brief decrease; if the acidification continues, the production is stimulated again. This second increase is first observed 33 days following the start of acidification but may have occurred before that.

In all cases where production is stimulated by acidification (Day 33 excepted), we observe an increase in photosynthetic specific activity denoting optimal metabolism of the algae. This observation contradicts the assertion in Ref *12*; but in these results, primary production data is expressed in disintegration per minute (DPM) without taking into account the ambiant CO_2 concentration; his conclusion in specific activity is, therefore, subject to caution.

The increase in specific activity is difficult to explain in relation to the changes brought by the acidification in the water chemical parameters. In fact, nutrients

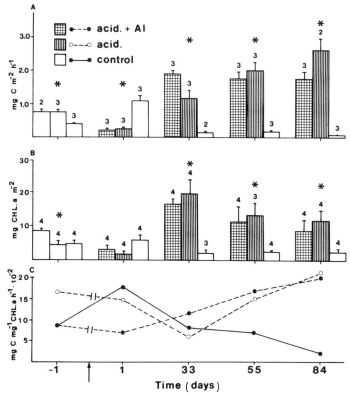

FIG. 3—*Continuous acidification, 3 months:* (A) *Primary production;* (B) *Chlorophyll-a;* (C) *Specific activity. Starting and ending of the acidification period are indicated by arrows. For A and B, X ± 1 standard error (SE), the number of replicates is indicated above the mean. Significant differences among conditions are indicated* *($p < 0.05$).

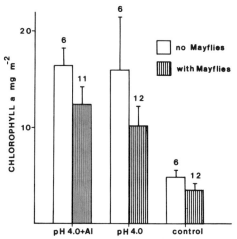

FIG. 4—*Predation assessment by measurement of chlorphyll-a.* X ± *1 standard error (SE), number of replicates above the mean.*

and microelements vary little and nonsignificantly after acidification. Potassium, however, does increase in the short-term acidification experiment; it is less than probable that this element could be responsible for this stimulation since it remains relatively constant in the case of the continuous acidification experiment, where there is a strong increase in periphytic production. One must keep in mind that these oligot[...]ents, which are often below[...]th this in mind, we cannot [...]s, or microelements, are [...]ion and readily absorbed by[...]detected. Also, added sulfu[...]observed stimulation. Th[...]or the activity of certain e[...]brane exchange mechanism[...]

The two [...]nagine a mechanism susce[...] another would bring the a[...]nce limits. For example, th[...]ification could, at short-ter[...]ns would lower the pH acc[...]

These op[...]decrease in the observed sh[...]ould disappear, being unab[...]rm stimulation observed on Day 33 would therefore correspond to the arrival of new species adapted to new pH conditions and capable of responding permanently. In fact, we notice that the initial periphyton community in both experimental setups were dominated by blue-greens and diatoms. In short-term acidification, after 12 h of treatment, *Pleurocapsa minor* became significantly more abundant[4]. In long-term acidification, the community shifted to green algae *Mougeotia*.

Biomass increase could be explained by a higher production due to a higher specific activity but also by a decrease of benthos grazing [28]. Several authors [13,16,29] have described the sensitivity of certain grazers to the lowering of pH. We have observed [20] in the same experimental setup that benthic invertebrates (mainly chironomids and mayflies) previously seen on top of the substrates, disappeared with experimental acidification. However, the analyses of the grazer exclusion experiment do not provide statistically significant results with respect to chlorophyll-a data even though it was always in greater quantity in the absence of grazers. Microinvertebrates were dominated by the rhizopoda: *Difflugia pristis* and *D. pulex*. Both were highly affected by acidification[5]. We do not know specifically the food regime of these two species; the only rhizopode specie present in our system that grazed algae was *Centropyxis discoides* [30], which repre-

[4] Caumartin, J., personal communication, Université du Quebec á Montreal, Montreal, Quebec.
[5] Costan, G. and Planas, D., "Effects of a Short-Term Experimental Acidification on a Microinvertebrate Community (Rhizopoda, Testacea)," *Canadian Journal of Zoology*, in press.

sented a low percentage of the community and apparently was not affected by acidification.[6]

Environmental conditions, mainly nutrient concentrations, also seem capable of modulating the stimulating effects. This is seen, at least in short term, since, in the experiment where phosphorus is nondetectable, no stimulation of the production was observed during the experiment while the biomass decreased in the first 24 h and stayed at a low level until the end of the experiment. Similar results showing that phosphorus is more important than H^+ ions in acidic ecosystems have been reported [*31–35*]. Such a difference in response with respect to environmental conditions could explain why the authors mentioned previously noted an increase in production in acidified milieu, whereas others observed the opposite or no reaction at all [*6–9,36*].

The experimental addition of aluminum had no effect on biomass or production even though observed concentrations exceeded levels mentioned as toxic by various authors [*37–39*] for the upper trophic levels. Here, again, chemical characteristics of water may explain this divergence in results. Aluminum toxicity varies greatly depending on its speciation and its capacity to form complexes with other constituents, in particular organic compounds [*40–42*]. In our study, the amount of organic carbon leads us to believe that the majority of aluminum must be complexed and thus is unavailable biologically.

Whatever the nature of the mechanisms involved, which are still hypothetical, it is possible to outline the ecological role of the two types of effects of acid precipitation: short-term acidification impact and continuous acidification impact.

The intermittent decreases in pH appear to have little effect on periphyton since in general the impact disappears 24 h after the pH has returned to its original values. However, if these intermittent decreases in pH are recurrent, or if an ecosystem is unable to recover between two precipitation events because of its low buffering capacity, the system is engaged in the continuous acidification process that has noticeable consequences on biomass and periphytic production. This latter response to continuous acidification does not appear to be negative since it leads to an increase in production. Our results, as those of Schindler [*10*], contradict the hypothesis coupling acidification with the process of oligotrophization stated by Grahn et al. [*43*].

To further appreciate the role of this increase in periphytic production in the ecology of lotic ecosystems, it remains to be seen if the algae responsible for this higher production are available to other trophic levels and if the usual grazers are still present following acidification.

Acknowledgments

Financial support was provided by Natural Sciences and Engineering Research Council of Canada, Ministère de l'Education du Québec (FCAC), and Canadian National Sportsmen's Fund. We wish to thank C. Gascon and J. Huot for the review of the manuscript.

[6] Costan, G., personal communication, Université du Quebec, á Montreal, Montreal, Quebec.

References

[1] Henriksen, A., in *Acid Rain/Fisheries*, R. E. Johnson, Ed., American Fisheries Society, Bethesda, Maryland, 1982, pp. 103–124.
[2] Haines, T. A., *Transaction of American Fisheries Society*, Vol. 110, No. 6, 1981, pp. 669–707.
[3] Hendrey, G. R., Baalstrud, K., Traaen, T. S., Laake, M., and Raddum, G., *Ambio*, Vol. 5, No. 5–6, 1976, pp. 224–227.
[4] Conway, H. L. and Hendrey, G. R. in *Acid Precipitation Effect on Ecological Systems*, F. M. D'Itri, Ed., Ann Arbor Sciences Publishers, Ann Arbor, MI, 1982, pp. 277–295.
[5] Hendrey, G. R. in *Acid Rain/Fisheries*, R. E. Johnson, Ed., American Fisheries Society, Bethesda, Maryland, 1982, pp. 125–135.
[6] Johnson, M. G., Michalski, F. P., and Christie, A. E., *Journal of Fisheries Research Board Canada*, Vol. 27, No. 3, 1970, pp. 425–444.
[7] Kwiatkowski, R. E. and Roff, J. C., *Canadian Journal of Botany*, Vol. 54, No. 22, 1976, pp. 2546–2561.
[8] Grahn, O., Hulteberg, H., and Landner, L. *Ambio*, Vol. 3, No. 2, 1974, pp. 93–94.
[9] Muller, P., *Canadian Journal of Aquatic Sciences*, Vol. 37, No. 3, 1980, pp. 355–363.
[10] Schindler, D. W. in *Proceedings International Conference Ecological Impact of Acid Precipitation*, D. Drablos and A. Tollan, Eds., SNSF Project, Norway, 1980, pp. 370–374.
[11] Stokes, P. M. in *Effects of Acidic Precipitation on Benthos*, R. Singer, Ed., North American Benthological Society, Springfield, Ill., 1981, pp. 119–138.
[12] Hendrey, G., "Effects of pH on the Growth of Periphytic Algae in Artificial Stream Channels," in *Acid Precipitation, Effects on Forest and Fish Project*, Report I R 25/76, Aas, Norway, 1976, p. 50.
[13] Hall, R. J., Likens, G. E., Fiance, S. B., and Hendrey, G. R., *Ecology*, Vol. 61, No. 4, 1980, pp. 976–989.
[14] Schindler, D. W., Wagemann, R., and Cook, R. B., *Canadian Journal of Aquatic Sciences*, Vol. 37, No. 3, 1980, pp. 342–354.
[15] Schindler, D. W. and Turner, M. A., *Water Air and Soil Pollution*, Vol. 18, No. 4, 1982, pp. 259–271.
[16] Arnold, D. E., Bender, P. M., Hale, A. B., and Light, R. W. in *Effects of Acidic Precipitation on Benthos*, R. Singer, Ed., North American Benthological Society, Springfield, Illinois, 1981, pp. 15–33.
[17] Serodes, J. B., Moreau, G., and Allard, M., *Water Research*, Vol. 18, No. 1, pp. 95–101.
[18] Gagnon, R. M., *Climat estival du Parc des Laurentides*, Ministère des Richesses Naturelles, Québec, Publication MP 35, 1970, p. 38.
[19] Ministère de l'Environnement du Québec, nonpublished data.
[20] Allard, M. and Moreau, G., "Effects of Experimental Acidification on a Lotic Macroinvertebrate Community," *Hydrobiologia*, in press.
[21] Vollenweider, R. A., *A Manual on Methods for Measuring Primary Production in Aquatic Environment*, International Biological Project (IBP) Handbook 13, Blackwell Scientific, Oxford, 1969.
[22] Schindler, D. W., *Nature* (London), Vol. 211, No. 5051, 1966, pp. 844–845.
[23] Marker, A. F. H., *Archiv fur Hydrobiologie Beihefte Ergebnisse der Limnologie*, Vol. 16, 1980, pp. 88–90.
[24] Marker, A. F. H., Crowther, C. A., and Gunn, R. J., *Archiv fur Hydrobiologie Beihefte Ergebnisse der Limnologie*, Vol. 16, 1980, pp. 52–62.
[25] Olsson, H., *Hydrobiologia*, Vol. 101, No. 1–2, 1983, pp. 49–58.
[26] Fromm, P. O., *Environmental Biology of Fish*, Vol. 5, No. 1, 1980, pp. 79–93.
[27] Wood, C. M. and McDonald, D. G. in *Acid Rain/Fisheries*, R. E. Johnson, Ed., American Fisheries Society, Bethesda, Maryland, 1982, pp. 197–226.
[28] Lamberti, G. A. and Resh, V. H., *Ecology*, Vol. 64, No. 5, 1983, pp. 1124–1135.
[29] Friberg, F., Otto, C., and Svensson, B. S. in *Proceedings International Conference Ecological Impact of Acid Precipitation*, D. Drablos, and A. Tollan, Eds., SNSF Project, Norway, 1980, pp. 304–305.
[30] Chardez, D., *Bulletin Institut Agronomique—Stations de Recherches*, Vol. 32, 1964, pp. 305–308.
[31] Hornstrom, E., Ekstrom, C., Millel, C., and Dickson, W., *Information from Sotvattens-Laboratoriet, Drottninghalm*, No. 4, 1973, p. 81.

[32] Yan, N. C., *Water, Air and Soil Pollution*, Vol. 11, No. 1., 1979, pp. 43–55.
[33] DeCosta, J. and Preston, C., *Hydrobiologia*, Vol. 70, No. 1, 1980, pp. 39–49.
[34] Shellito, G. A. and DeCosta, J., *Water, Air and Soil Pollution*, Vol. 16, No. 4, 1981, pp. 415–431.
[35] Yan, N. C., Lafrance, C. J., and Hitchin, G. G. in *Acid Rain/Fisheries*, R. E. Johnson, Ed., American Fisheries Society, Bethesda, Maryland, 1982, pp. 137–154.
[36] Conroy, N., Hawley, K., Keller, W., and Lafrance, C. J. in *Proceedings First Specialty Symposium on Atmospheric Contribution to the Chemistry of Lake Waters*, International Association Great Lake Research, 1975, pp. 146–165.
[37] Schofield, C. L. and Trojnar, J. R. in *Polluted Rain*, T. Y. Toribara, M. W. Miller, and P. E. Morow, Eds., Plenum Publication Corp., New York, 1980, pp. 341–365.
[38] Dickson, W., *Verhein International Vereinigung Limnologie*, Vol. 20, 1978, pp. 851–856.
[39] Muniz, I. P. and Leivestad, H. in *Proceedings International Conference Ecological Impact of Acid Precipitation*, D. Drablos and A. Tollan, Eds., SNSF project, Norway, 1980, pp. 320–321.
[40] Driscoll, C., Baker, J. P., Bisogni, J. J., and Schofield, C. L., *Nature*, Vol. 284, No. 5752, 1980, pp. 161–163.
[41] Baker, J. and Schofield, C. L. in *Proceedings International Conference Ecological Impact of Acid Rain Precipitation*, D. Drablos and A. Tollan, Eds., SNSF project, Norway, 1980.
[42] Baker, J., Ph.D. thesis, Cornell University, Ithaca, NY, 1981, p. 441.
[43] Grahn, O. H., Hulterberg, H., and Landner, L., *Ambio*, Vol. 3, No. 2, 1974, pp. 93–94.
[44] Longpré, G., Joubert, G., and Trottier, J., "Guide d'Information sur l'Analyse Physique, Chimique, Biologique et Bactériologique des Milieux Environnementaux," Ministère de l'Environnement du Québec, Direction Générale des Inventaires et de la Recherche, Direction Générale des Laboratoires, Québec, 1982, p. 149.
[45] Mackereth, F. J. H., Heron, J., and Talling, J., *Water Analysis*, Freshwater Biological Association, Publication No. 36, 1978, p. 120.
[46] Stainton, M. P., Capel, M. J., and Armstrong, F. A. J., *The Chemical Analysis of Freshwater*, Fisheries Research Board of Canada, Miscellaneous Special Publication No. 25, 1974, p. 125.

D. C. L. Lam,[1] A. G. Bobba,[2] D. S. Jeffries,[3] and J. M. R. Kelso[4]

Relationships of Spatial Gradients of Primary Production, Buffering Capacity, and Hydrology in Turkey Lakes Watershed

REFERENCE: Lam, D. C. L., Bobba, A. G., Jeffries, D. S., and Kelso, J. M. R., "**Relationships of Spatial Gradients of Primary Production, Buffering Capacity, and Hydrology in Turkey Lakes Watershed,**" *Impact of Acid Rain and Deposition on Aquatic Biological Systems, ASTM STP 928,* B. G. Isom, S. D. Dennis, and J. M. Bates, Eds., American Society for Testing and Materials, Philadelphia, 1986, pp. 42–53.

ABSTRACT: The carbon uptake rate of phytoplankton was found to increase from poorly buffered, headwater lakes to better buffered, downstream lakes in the Turkey Lakes Watershed near Sault Ste. Marie, Ontario, Canada. While the nutrient, sunlight, and temperature conditions remained fairly uniform, the pH, alkalinity, and dissolved inorganic carbon concentration increased in a similar manner as the primary production. Discussions are presented on the relationships of these spatial gradients and their probable causes in such a small watershed (area = 10.5 km^2). In particular, results from a hydrological model show that more groundwater flowed into downstream lakes than headwater lakes. The greater soil-water contact time and the higher concentration of calcium carbonate ($CaCO_3$) present in the till at low-lying areas cause the introduction of relatively greater quantities of Ca^{2+} and alkalinity into the downstream lakes. Thus, the development of acidification-eutrophication models must incorporate the physical, chemical, and biological processes for both soil and water.

KEY WORDS: acidification, carbon uptake, buffering capacity, algal growth, hydrological model, watershed

While a number of studies have demonstrated the effects of acidification on phytoplankton [1–3], the underlying causes, particularly those affecting primary production, have not been discussed adequately. Apart from being at the primary level of the aquatic food chain, algae utilize chemical nutrients and contribute directly to the carbon cycle through primary production and respiration. Thus, algal growth is tied closely with the water chemistry.

[1,3] Research scientists and [2] physical scientist, National Water Research Institute, Canada Centre for Inland Waters, Burlington, Ontario, Canada.
[4] Research scientist, Great Lakes Fisheries Research Branch, Sault Ste. Marie, Ontario, Canada.

The water chemistry, in turn, is controlled by the soil-water interactions [4]. Given the vast regional differences in soil buffering capacity and watershed hydrology, it is conceivable that algae in natural lakes within the same watershed could exhibit different growth rates, even if the sulfate deposition rate and the nutrient conditions are apparently the same. Indeed, by the same hypothesis, if the variability in buffering capacity and hydrology is large enough, differences in water chemistry and hence algal growth could appear in different locations of the same watershed (for example, headwater lakes versus downstream lakes).

The purpose of this paper is to present evidence from the Turkey Lakes Watershed in Canada apparently in support of such a hypothesis. Both the observed carbon uptake rate and the buffering capacity show spatial gradients from headwater to downstream lakes. Recent results from a hydrological model also show similar gradients in the groundwater flow in streams. It is, therefore, interesting to examine how these biological, chemical, and physical gradients relate to each other. In particular, we do not intend to rule out other hypotheses that also may explain these gradients. As we are in the process of developing a watershed acidification model, an examination of several hypotheses in the light of the available data is crucial at this stage. On the one hand, the investigation will help in deciding on the choice of model components and model complexity. On the other hand, it will identify possible gaps in the existing data sets and can be used to improve the planning of future experiments.

The Study Area

The Turkey Lakes Watershed (Fig. 1) is situated about 50 km north of Sault Ste. Marie, Ontario. The atmospheric acid deposition is moderate in this region, with an average precipitation pH of 4.3 to 4.5 and an annual sulfate load of 20 to 30 kg $SO_4^=$/ha. A distinctive feature of the watershed is that it contains a series of five lakes over a total watershed area of 10.5 km^2. While similar meteorological conditions may prevail over such a small area, the morphological and geological features vary substantially [5]. For example, the highest elevation is 645 m and the lowest, 245 m. The soil is mainly loam and sandy loam with a depth of 0.2 to 1 m at higher elevations and is mainly gravelled till and fine-grained till with a depth of 1 to 10 m at lower elevations. The cation exchange capacity is about 10 to 20 meq/100 g at higher elevations and is about 10 to 40 meq/100 g at lower elevations [6]. Thus, given such variations of soil depths and chemistry, the hydrological pathway and contact time of the major ions are expected to differ from location to location.

An intensive research project has been carried out at this watershed since 1979, with a network of strategically located sampling sites (Fig. 1) covering each of the five lakes (L1 to L5) and their inflows and outflows (S0 to S5). The study team consists of researchers from various scientific disciplines including the areas of atmosphere, soil, water, hydrology, forestry, and fisheries. The list of routinely sampled variables at this watershed is extensive [7–9]. Efforts have been made

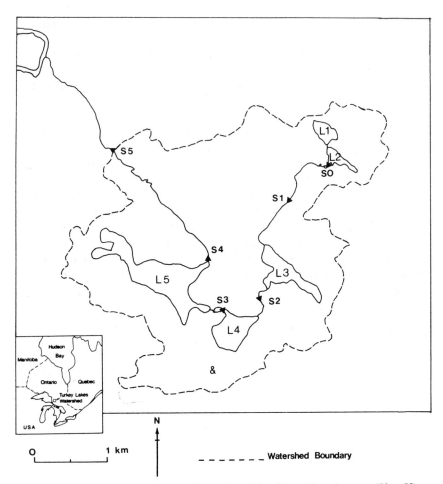

FIG. 1—*Turkey Lakes Watershed sampling stations: lakes (L1 to L5) and streams (S0 to S5).*

to integrate some of the research results through the development of acidification models [*10–11*].

Observed Data

Collins et al. [*9*] first noted the spatial differences in the observed carbon uptake rates of the five lakes (L1 to L5). These rates (Fig. 2a) were obtained from *in situ* samples taken by the ^{14}C dark bottle method with correction for the background dissolved inorganic carbon. In general, the primary productivity as shown by these rates is lower at the headwater lakes (L1 and L2) than the downstream lakes (L3, L4, and L5). At peak production period, the uptake rates can differ by two- to three-fold between L1 and L5. The dominant species are blue-green algae, which are commonly found in the pH range of 5 to 7 [*1*]. In particular, *Meris-*

mopedia punctata was the sole Cyanophyte in L1 and L2 during the summer period. In the other lakes, the same species was significant, but species shifts did occur with major contributions from *Chroococcus dispersus, Aphanothece clathrata,* and *Microcystis flos-aquae* to the summer bloom [9].

For the range of pH and water temperature commonly found in these lakes, the partial pressure of carbon dioxide, P_{CO_2}, can be conveniently expressed as [12]

$$-\log (P_{CO_2}) = pH - \log (HCO_3^-) - 7.8 \tag{1}$$

where the concentration of HCO_3^- is in mole/L and temperature correction is small. Using Eq 1 and the observed pH and bicarbonate data, we computed the content of carbon dioxide as log (P_{CO_2}) and plotted it (Fig. 2b). Not surprisingly, there was more carbon dioxide in headwater lakes than downstream lakes.

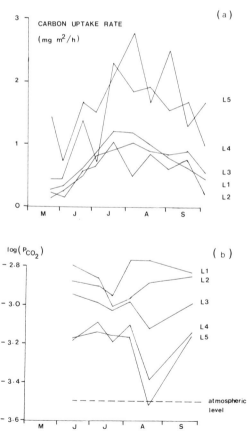

FIG. 2—(a) *Observed carbon uptake rates for L1 to L5, May to September, 1981 (see Ref 9);* (b) *dissolved carbon dioxide expressed as* log (P_{CO_2}) *(Eq. 1) for L1 to L5, May–Sept., 1981.*

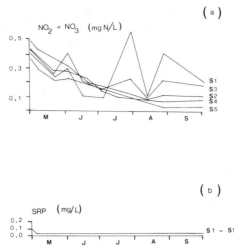

FIG. 3—(a) Observed nitrite plus nitrate concentration, and (b) observed soluble reactive phosphorus concentrations, for May–Sept., 1981, at S1 to S5.

By contrast, there appears to be no particular spatial gradient or time pattern recognizable in the nutrient loading data. For example, Figure 3a shows the observed nitrite plus nitrate (NO_2^- + NO_3^-), and Figure 3b shows the observed soluble reactive phosphorus (SRP) for stations S1 to S5 for 1981. The NO_2^- + NO_3^- is measured by the autoanalyzer colorimetric method using cadmium reduction, and the SRP is the molybdate reactive orthophosphate measured also by using an autoanalyzer colorimetric procedure. Note that the stations shown in Figures 3a and 3b are either at the inflow or at the outflow of the lakes and are therefore representative of the loading conditions at these sites. For example, S2 data can be used to assess the input loading to L4 (see Fig. 1). As shown in these two figures, there are no statistically significant differences among the inflow nutrient concentrations, particularly the SRP.

Besides the inputs from the streams, the inputs from nonpoint sources and runoffs from the watershed also contribute to the nutrient loading. Since the entire

TABLE 1—Watershed areas, lake areas, and water renewal times (= lake volume/outflow).

	Watershed Area, ha	Lake Area, ha	Lake Area/ Total Area	Water Renewal Time, Yr
L1	24.0	5.88	0.245	1.30
L2	61.7	5.82	0.0943	0.30
L3	337	19.2	0.0570	0.15
L4	491	19.2	0.0391	0.25
L5	803	52.0	0.0648	0.94

NOTE: 1 ha = 10 000 m^2.

watershed is forested with no major industrial or anthropogenic nutrient sources, these nonpoint sources can be approximately represented by the ratio of the lake area to the watershed area for each lake. As shown in Table 1, this ratio decreases between L1 and L4 and then increases for L5. In spite of the substantial drop between the ratios for L1 and L2, the phytoplankton productivities do not show a similar decrease (Fig. 2a). It appears therefore that the loading inputs from both streams and nonpoint sources are too low to influence the biological productivities significantly. Furthermore, the water renewal time of the five lakes does not show any obvious spatial gradient at all (Table 1). In fact, the most upstream lake (L1) and the most downstream lake (L5) have longer renewal times, whereas the middle lake (L3) has the shortest. Thus, many of these conventional indicators of nutrient-plankton relationships cannot satisfactorily explain the observed spatial gradient in the biological productivities in these lakes.

It is possible that other better limnological indicators may provide greater scientific insight into explaining the observed phenomenon. For example, the phosphorus turnover rates, the ratio between seston nitrogen and phosphorus and other factors affecting the nutrient uptake activities may affect the primary production. Unfortunately, we did not measure these rates and ratios. On the other hand, we did measure the sunlight and water temperature which can affect these kinetic rates, but they are found to be similar for all the five lakes [9]. It is, therefore, interesting to find if other limnological observations show similar gradients.

Since the Turkey Lakes are oligotrophic, with low nutrient inputs and fast water renewal times, they are more susceptible to excessive sulfate inputs than eutrophic lakes [13,1]. In particular, the pH and alkalinity levels will be affected by the sulfate load. Figure 4a shows the observed pH and Figure 4b, the observed alkalinity. The pH is measured with glass electrodes and the alkalinity with Metrohm tetraprocessor employing end point detection equivalent to the Gran technique. The pH increases by almost one unit between S1 and S5 and the alkalinity by fourfold. In both cases, the increase is progressive, with higher pH and alkalinity as we go downstream. Even in August and September, when the pH and alkalinity levels in all streams appear to be maximum, the spatial gradient is still maintained. The fact that the phytoplankton shows a species shift from upstream lakes to downstream lakes is also consistent with the pH and alkalinity spatial gradients. Hence, it is important to investigate the relationships among these chemicals and biological parameters following the hydrological pathways.

Discussions and Hydrological Model Results

There have been a number of studies on the effects of acidification on phytoplankton [13–16]. The effects include significant decreases in species diversity [2], toxicity of heavy metals [17], species shifts [1], and inhibition of certain protein carriers by bisulfite [18]. Some authors reported growth reduction as the pH decreased [19], but others concluded differently [20].

48 IMPACT OF ACID RAIN

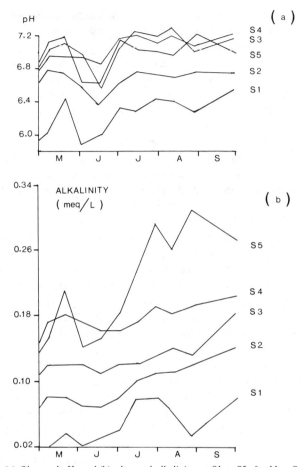

FIG. 4—(a) *Observed pH, and* (b) *observed alkalinity, at S1 to S5, for May–Sept., 1981.*

All these results are probably significant, but it is definitely difficult to compare them, because the methods, locations, and background conditions differ. It is not the intention of this paper to resolve these differences. Rather, we want to examine whether a study based on a natural system of progressively buffered lakes with apparently similar nutrient backgrounds such as the Turkey Lakes system would avoid most of these discrepancies and would manifest more clearly the acidification effects. While we do not object to manipulative experiments or laboratory studies as they definitely have merits in their own right, we do feel a natural system of lakes deserves a closer look.

Since the total area of the watershed is so small, the solar radiation, air temperature, and other meteorological conditions do not vary significantly from one lake to the other. Indeed, Collins et al. [9] reported that the surface water temperatures of the five lakes were similar, except for the shallowest lake, L3, which was slightly warmer than the others. As shown in Fig. 3, the nutrient loadings, particularly the phosphorus inputs, flowing into and out of these lakes did not exhibit much spatial difference, either. If anything, nutrient data from Fig. 3 would suggest that algal production would be greater in the headwater lakes. Thus, it appears that the major factors governing photosynthesis, such as light and nutrients, were fairly spatially uniform in this watershed.

By contrast, the carbon uptake rate of the phytoplankton showed a two- to three-fold increase from the headwater lakes to the downstream lakes (Fig. 2a). Clearly, this phenomenon cannot be explained by the observed nutrient and sunlight conditions. It must be more related to the pH, CO_2, and alkalinity, all showing gradients similar to the primary productivity gradient.

In the case of dissolved carbon dioxide (Fig. 2b), L1 showed a partial pressure consistently higher than the atmospheric equilibrium CO_2 pressure [$-\log (P_{CO_2}) = 3.5$], but L5 showed a CO_2 level very close to the atmospheric level, particularly during the midsummer peak algal production period. Thus, it appears that the more productive lakes at lower elevations consumed more carbon dioxide. This deficit of carbon dioxide is in agreement with the theoretical findings of Kelly et al. [21]. These authors used a mass balance model incorporating the photosynthesis and carbon equilibrium equations to show that phytoplankton production is able to reduce the dissolved carbon dioxide to a rather low level, even at normal reaeration and alkalinity levels. However, whether such a low level of CO_2 can affect algal growth has long been a controversial question [22,23]. In the present case, the algal growth did not appear to be affected by the CO_2 deficit, since there was more production in lakes with less dissolved CO_2. Indeed, the average dissolved inorganic carbon concentrations for the five lakes, L1 to L5, are 1.6, 2.3, 2.6, 3.1, and 3.3 mg C/L, respectively, and are, therefore, more in line with the observed gradient of increasing primary productivity than the CO_2 partial pressures.

The dissolved inorganic carbon concentration, in turn, is directly related to the alkalinity, as most of it is probably HCO_3^- for the range of pH found in these lakes. As we have seen in Fig. 4, the observed spatial gradient of alkalinity is remarkably prominent, and so is the pH gradient. Baker [19] found that, for pH ranging from 5 to 7, reduction in pH generally decreased the algal growth and increased the permeability of the cell membrane, causing the disruption of the gradient maintained between the intracellular and the extracellular concentrations of certain protein carriers. Since the toxicity due to heavy metals does not play a strong role at this pH range and the influences of nutrient, sunlight, and dissolved carbon dioxide have just been ruled out, the finding by Baker [19], at the present stage, seems to provide one of the more acceptable reasons that could explain the primary production gradient by reference to the water chemistry gradients. However, more research is needed to confirm the findings.

As the occurrence of the biological carbon uptake rate gradient is attributable to the water chemistry gradients, the occurrence of the water chemistry gradients themselves can be related to the physical hydrology of the watershed. Figures 5a

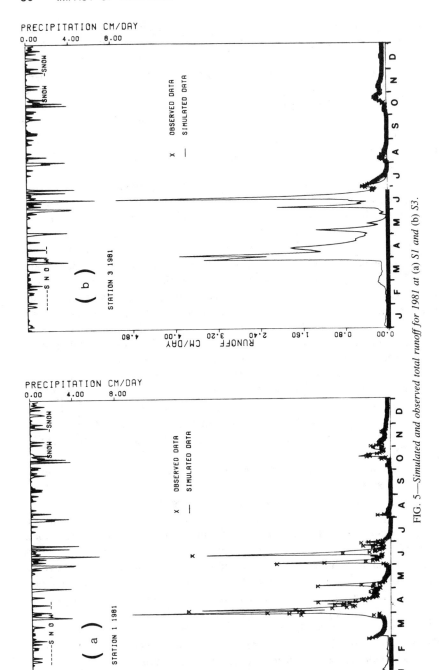

FIG. 5—Simulated and observed total runoff for 1981 at (a) S1 and (b) S3.

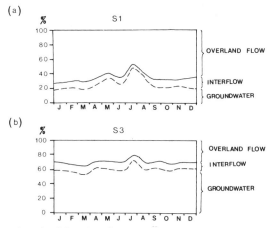

FIG. 6—*Computed overland flow, interflow, and groundwater flows expressed as monthly mean percentages of the total runoff at* (a) *S1 and* (b) *S3.*

and 5*b* show the total watershed runoff computed by a hydrological model [*11*] for S1 and S3, respectively. This model has been verified with the 1981 and 1982 data obtained at S1 to S5, and the agreement between computed and observed results is good [*10*]. Based on this model, we also computed the groundwater flow into the streams. This computed groundwater flow has been verified with ^{18}O isotope data and has been found to be fairly accurate [*10*]. Figures 6*a* and 6*b* show the monthly average overland flow, interflow, and groundwater flow expressed as percentages of the total runoff. Clearly, the headwater area (S1) received less groundwater (about 30% of total runoff) than the downstream area (S3), which received about 60% of its runoff from groundwater flow. Since the soil layer is deeper and contains more $CaCO_3$ at downstream areas [*6*], the high proportion of groundwater flow will convey more Ca^{2+} into the downstream lakes. Thus, as we have seen, this creates the observed spatial gradients not only for Ca^{2+} and alkalinity, but also for pH and, subsequently, primary production.

Conclusions

In summary, observations showed that the carbon uptake rate of the phytoplankton increases from poorly buffered, first order lakes to better buffered, downstream lakes in the Turkey Lakes Watershed. While the effects of nutrients, sunlight, and temperature are quite similar in the watershed, the levels of pH, alkalinity, and dissolved inorganic carbon increased progressively downstream as did the carbon uptake rate. This phenomenon leads to the conclusion that the algal growth is apparently more influenced by the latter factors than the former in this watershed. Hydrological model results show that more groundwater enters the low-lying streams than enters headwater streams. The results suggest that because of the greater contact time, more Ca^{2+} must have entered the downstream areas and will eventually cause the development of both the chemical and the biological gradients. However, this hypothesis should be confirmed by further conducting experiments to measure the phosphorus turnover rate, the seston phosphorus to nitrogen ratio, and other nutrient flux rates.

The development of models relating acidification to eutrophication is a complex undertaking. On the one hand, watershed acidification encompasses the physical, chemical, and biological processes in soil and water and therefore requires a moderately complex model structure [10]. The hydrological model discussed here is an example of a part of such a framework. On the other hand, other studies have shown that, subjected to different sites and background conditions, the pH may affect the plankton slightly, may cause acute toxicity, or, as in the present case, may exert a moderate influence. In terms of developing acidification-eutrophication models, therefore, the incorporation of research results from laboratories and experimental and natural lakes must take into account such background differences.

Acknowledgments

The authors thank C. K. Minns, P. T. S. Wong, M. E. Thompson, F. C. Elder, R. G. Semkin, W. A. Glooschenko, and P. J. Dillon for helpful discussions.

References

[1] Yan, N. D. and Stokes, P. M., *Proceedings*, 11th Canadian symposium on Water Pollution Research, 1976, Jones, P. H., Ed., University of Toronto, Ontario, pp. 127–137.
[2] Kwiatkawski, R. E. and Roff, J. C., *Canadian Journal of Botany*, Vol. 54, 1976, pp. 2546–2561.
[3] Hendrey, G. R., Kaalsrud, K., Traan, T. S., Laake, M., and Raddum, A., Vol. 5, 1976, pp. 224–227.
[4] Likens, G. E., Wright, R. F., Galloway, J. N., and Butler, T. J., *Scientific America*, Vol. 241, 1979, p. 43.
[5] Jeffries, D. S. and Semkin, R. G., "Basin Description and Information Pertinent to Mass Balance Studies of the Turkey Lakes Watershed," Report TLW-82-01, Canada Centre for Inland Waters, Burlington, Ontario, Canada, 1982.
[6] Cowell, D. W. and Wickware, G. M., "Preliminary Analyses of Soil Chemical and Physical Properties, Turkey Lakes Watershed, Algoma, Ontario," Report TLW-83-08, Canada Centre for Inland Waters, Burlington, Ontario, Canada, 1983.
[7] Jeffries, D. S. and Semkin, R. G., "Data report: Major Ion Composition of Lake Outflows and Major Streams in the Turkey Lakes Watershed (January, 1980–May, 1982)," Report TLW-83-05, Canada Centre for Inland Waters, Burlington, Ontario, Canada, 1983.
[8] Foster, N. W., *Forest Ecology and Management*, Vol. 12, 1985, pp. 215–231.
[9] Collins, R. H., Love, R. J., Kelso, J. R. M., Lipsit, J. H., and Moore, J. E., "Phytoplankton Production, as Estimated by the ^{14}C Technique, and Populations Contributing Production 1980/1981, in the Turkey Lakes, Ontario, Watershed," Canadian Technical Report of Fisheries and Aquatic Sciences, No. 1191, Canada Centre for Inland Waters, Burlington, Ontario, 1983.
[10] Lam, D. C. L. and Bobba, A. G., "Modelling Watershed Runoffs and Basin Acidification," contributions to *Proceedings*, UNESCO-IHP Workshop on Hydrological and Hydrogeochemical Mechanisms and Model Approaches to the Acidification of Ecological Systems, Uppsala University, Uppsala, Sweden, 1985, NHP–No. 10, pp. 205–215.
[11] Bobba, A. G. and Lam, D. C. L., "Application of Linearly Distributed Surface Runoff Model for Watershed Acidification Problems" in *Proceedings*, Canadian hydrology symposium, University of Laval, Quebec, Canada, 11–12 June, 1984, in print.

[12] Thompson, M. E. and Hutton, M. B., "Sulfate in Lakes of Eastern Canada: Calculated Atmospheric Loads Compared with Measured Wet Deposit," NWRI report, Canada Centre for Inland Waters, Burlington, Ontario, 1982.
[13] Grahn, O. H. and Hultberg, H., "Effects of Acidification on the Ecosystem of Oligotrophic Lakes—Integrated Changes in Species Composition and Dynamics," Inst. Vatten-och Luftvardsforskning, Gothenburg, Meddelande nr.2, Sweden, 1974.
[14] Wodzinski, R. S. and Alexander, M., *Journal Environmental Quality*, Vol. 7, 1978, pp. 358–360.
[15] Grahn, O. H., Hultberg, H., and Landner, L., *Ambio*, Vol. 3, 1974, pp. 93–94.
[16] Dillon, P. J., Jeffries, D. S., Snyder, W., Reid, R., Yan, N. D., Evans, D., Moss, J., and Scheider, W. A., *Journal Fisheries Research Board Canada*, Vol. 35, 1978, pp. 809–815.
[17] Baker, M. D., Mayfield, C. I., Innis, W. E., and Wong, P. T. S., *Chemosphere*, Vol. 12, 1983, pp. 493–497.
[18] Wong, P. T. S., Chau, Y. K., and Patel, D., *Chemosphere*, Vol. 11, 1982, p. 367.
[19] Baker, M. D., "Effects of Acidification, Metals, Metalloids and Bisulfite on Aquatic Microorganisms," Ph.D. thesis, Department of Biology, University of Guelph, Guelph, Ontario, Canada, 1983.
[20] Schindler, D., Wagemann, R., Cook, R., Ruscynski, T., and Prokopovich, J., *Canadian Journal Fisheries and Aquatic Sciences*, Vol. 37, 1980, pp. 342–354.
[21] Kelly, M. G., Church, M. R., and Hornberger, G. M., *Water Resources Research*, Vol. 10, 1974, pp. 493–497.
[22] King, D. L., *Journal Water Pollution Control Federation*, Vol. 42, 1970, pp. 2035–2051.
[23] Goldman, J. C., Porcella, D. B., Middlebrooks, E. J., and Toerien, D. F., *Water Research*, Vol. 6, 1972, pp. 637–679.

Jon W. Allan[1] and Thomas M. Burton[2]

Size-Dependent Sensitivity of Three Species of Stream Invertebrates to pH Depression

REFERENCE: Allan, J. W. and Burton, T. M., **"Size-Dependent Sensitivity of Three Species of Stream Invertebrates to pH Depression,"** *Impact of Acid Rain and Deposition on Aquatic Biological Systems, ASTM STP 928,* B. G. Isom, S. D. Dennis, and J. M. Bates, Eds., American Society for Testing and Materials, Philadelphia, 1986, pp. 54–66.

ABSTRACT: Three species of stream invertebrates, the caddis fly *Lepidostoma liba* (Ross), the isopod *Asellus intermedius* (Forbes), and the snail *Physella heterostropha* (Say), were found to be highly vulnerable to depression to pH 4.0 with 70% sulfuric acid (H_2SO_3) and 30% nitric acid (HNO_3) from a control pH of 6.7 to 7.2. This vulnerability was directly correlated to size with smaller individuals being much less tolerant to pH 4 depression than were larger individuals. Smaller size classes, including egg masses of *P. heterostropha*, were totally eliminated by exposure to pH 4. Survival increased as initial body mass increased past a threshold size (0.5 mg dry weight per individual for *A. intermedius*, 0.6 mg for *L. liba*, and at a shell length of 4 mm for *P. heterostropha*). Depression of pH from 7 to 4 also depressed growth rates for *A. intermedius*, and *L. liba*. This experiment was conducted in recirculating laboratory streams operated near ambient temperatures at 8-week intervals throughout the year.

KEY WORDS: pH effects, stream invertebrates, mortality, growth, size dependency

Acidification of aquatic ecosystems to pH 5 or less can cause mortality for many species of invertebrates [1–9]. Other acidification effects include decreased emergence [6], increased drift [6,10], and increased food consumption [9] but decreased growth [11]. There is some evidence that such effects vary with size and maturity of invertebrates [11]. Until this study, this size-dependent vulnerability has received limited attention.

Materials and Methods

Three species of invertebrates selected for intensive study included a caddis fly, *Lepidostoma liba* (Ross), an isopod, *Asellus intermedius* (Forbes), and a snail,

---

[1] Graduate research assistant, Department of Zoology, Michigan State University, East Lansing, MI 48824.
[2] Professor, Departments of Zoology and Fisheries and Wildlife, Michigan State University, East Lansing, MI 48824.

Physella heterostropha (Say). Taxonomic references used for identification and nomenclature of these species included Merritt and Cummins [*12*], Wiggins [*13*], and Ross [*14*] for *L. liba;* Pennak [*15*] and Williams [*16*] for *A. intermedius;* and Harman and Berg [*17*] and Burch [*18*] for *P. heterostropha*. *Physella* was accepted as the genus for *P. heterostropha* [*18*] rather than *Physa* as had been used in earlier publications [*4,19*]. No confirmations of species identifications have been obtained, and identification is solely that of the authors.

Specimens of the three selected species were collected at intervals throughout the year from low alkalinity (160 to 300 µeq/L) streams near Paradise, Chippewa County, Michigan. These specimens were placed in a pair of laboratory streams at Michigan State University.

Each of the 234-cm-long by 56-cm-wide by 28-cm-deep recirculating laboratory stream channels consisted of a wooden trough with a center divider. A paddle wheel circulated water maintained at a depth of 16 cm down one 28-cm side of the trough, through an opening in the divider, around the other side, and through an opening at the opposite end of the divider back to the paddle wheel. The paddle wheels in both streams were the same size and driven by the same motor so that circulation in the two channels was similar. The pine wood was painted with epoxy paint and caulked with silicon sealant to prevent leakage. Stream water obtained from the streams where the invertebrates were collected was used to fill the laboratory streams. Evaporated stream water was replaced with deionized water to maintain depth and prevent concentration of dissolved constituents. Temperature in these laboratory streams was partially cooled on hot summer days by circulating well water through coils of flexible, plastic tubing in the channels. Temperatures were 3 to 5°C lower than ambient air temperature on hot summer days and were maintained just above freezing in the winter (2 to 3°C) but otherwise varied with ambient temperatures (2 to 22°C depending on the season).

One of the laboratory streams was maintained as a control with the pH varying from 6.7 to 7.2, similar to conditions in the streams where the invertebrates were collected. The other channel was acidified to pH 4.0 by slowly dripping in a 0.2 N solution of 70% sulfuric acid (H_2SO_4) and 30% nitric acid (HNO_3) over a period of several days. This acidified stream was maintained at pH 4 and did not vary more than 0.2 pH units during the rest of the experiment.

Ten similar-sized individuals of each of the three species were placed in cages in the streams (three replicates of ten individuals per replicate per species). The 15-cm by 10-cm by 6-cm cages were constructed of 1-mm mesh nylon bolting cloth stretched over a plexiglass frame.

After complete mortality in any channel or at 8-week intervals (whichever came first), previous experimental animals were discarded and ten new individuals for each replicate for each species were placed in the cages in each stream. All specimens were acclimated to laboratory conditions in the reference stream for at least 2 weeks before being subsampled for use in the cages. After placing the individuals in cages, exposure to acidified stream water (pH 4) was immediate with no acclimation period.

All three species are capable of living on leaf detritus [*12,15*]. *L. liba* is a leaf-shredding detritivore [*12*]; *A. intermedius* is a scavenger or general detritivore known to ingest dead plant material [*15*]; and *P. heterostropha* is an algal grazer but also frequently ingests dead plant material [*15*]. Thus, 5 g of dry white birch leaves (*Betula papyrifera*) were preconditioned by allowing a 30-day microbial and algal colonization period in stream water and were placed in the cages as a food source. These freshly fallen leaves were collected from the forest floor adjacent to the streams near Paradise, Michigan at the time of autumnal leaf fall and were dried at 60°C and stored until used. Preconditioned leaf packs were replaced with newly colonized leaves at 7-day intervals to reduce the possibility of food limitation due to acidification-induced decreases in microbial populations.

At intervals of 7 days or less, the cages were removed from the water and all specimens were located and carefully examined for mortality. Mortality of snails was usually obvious due to lack of attachment and shell discoloration (shells turned very pale compared to live specimens). In addition, snails were deemed dead when there was no response to prodding the operculum and when the snail failed to initiate movement after being placed back on a substrate for more than 2 to 5 min of observation. Mortality was obvious for *Asellus* due to lack of activity. Live specimens quickly moved, seeking cover when disturbed. *Lepidostoma* resisted extraction from cases and tended to move quickly when placed on substrates, so mortality could be determined easily.

Differences in survival in the control and acidified streams at each interval were tested using students t tests with actual numbers surviving at each interval. For ease of graphic display, these numbers were converted to per cent survival after statistical treatment and are presented in that manner in the results section. Since actual numbers were used in the statistical procedures, no arc sine $\sqrt{n}$ transformations were necessary. To compare effect of body size on survival, regressions were developed and differences in response between control and acidified streams were tested using the f statistic as discussed by Gill [*22*]. Differences were accepted as significant at the $p < 0.05$ level for all statistics unless otherwise indicated.

While the cages were removed from the water at each interval, all living *Asellus* and *Lepidostoma* were removed from the moist leaf surfaces, and length was measured from the anterior tip of the head to the posterior tip of the abdomen (not including appendages such as antennae or abdominal appendages). Length measurements were used with previously developed length-weight regressions for these two species to calculate dry weight. These calculated weights were used to determine instantaneous growth rates according to the formula used by Cummins et al. [*20*] and Grafius and Anderson [*21*] as follows:

$$\text{Growth Rate} = \frac{\text{Ln final weight (mg)} - \text{Ln initial weight (mg)}}{\text{time (days)}}$$

Since growth rate is known to be highly dependent on temperature, regressions of temperature versus growth rate for the control and acidified environments were compared for both *Lepidostoma* and *Asellus* using the f statistic [22].

Only mortality was recorded for *Physella*. Mortality for all three species was calculated as percent of original numbers remaining at the end of the test interval. After mortality and length determinations, all cages were placed back into the stream if mortality were not complete.

When no specimens in the acidified stream survived or after 8 weeks, a new test group of three replicates of ten individuals per replicate was introduced and monitored in the same manner. Six size classes of *Asellus* (0.20, 0.48, 0.63, 1.40, 2.25, and 4.31-mg mean dry wt/individual), eight of *Lepidostoma* (0.05, 0.075, 0.25, 0.35, 0.64, 1.00, 1.53, and 2.38-mg mean dry wt/individual), and three of *Physella* (4, 7, and 12 mm average length from tip of apex to outermost point on aperture) were tested. These size classes represented mean size classes of live individuals from collections from the streams near Paradise, Michigan in 1982 on July 24, Sept. 9, Oct. 25, Nov. 30, and from a Feb. 4, 1983, collection plus subsequent growth in the reference stream until time of testing. The growth in the laboratory streams is discussed in detail in the results that follow. Since laboratory temperatures simulated outdoor temperatures and since new specimens were obtained from outdoor streams at several intervals, the size of individuals tested in the laboratory at any season was near the mean size of specimens occurring in the natural streams.

The cages for each species were placed in similar positions in the control and acidified channels. Flow was not measured at each cage, but careful attention to position and orientation should have ensured that minimal differences in flow existed between the control and acidified channels.

Results

Asellus intermedius

For any size class of *Asellus* tested, mortality was significantly greater in the acidified stream than in the control stream for each interval tested (Fig. 1). However, the rate of mortality was highly correlated with body size (Fig. 1). All small size class individuals (0.20 ± 0.01-mg dry weight) died within four days of exposure to pH 4. As weight increased, some individuals were able to survive for longer intervals with more and more individuals surviving for the complete 8 weeks exposure as size increased to 1.40 ± 0.04-mg dry weight per individual (Fig. 1). This trend of decreased survival with decreases in body weight in the pH 4 treatments was statistically significant (f statistic, $p < 0.05$) as was the trend of increasing mortality with time for each size class tested when compared to control populations (f statistic, $p < 0.01$) (Fig. 1).

Those specimens that did survive in the pH 4 treatments (Fig. 1) grew at much slower rates than did their counterparts in the control stream (Fig. 2). Reduced growth rate was especially obvious for the smaller size classes but tended to

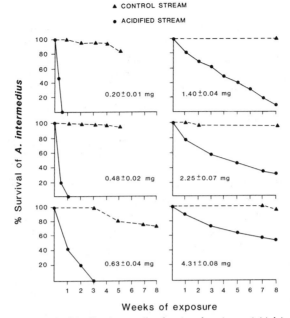

FIG. 1—*Percent survival of* Asellus intermedius *by size class (mean initial individual mg dry weight ± one standard deviation).*

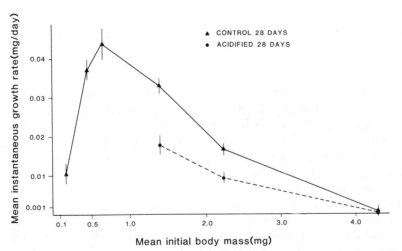

FIG. 2—*Mean instantaneous growth rate (mg/day) of* Asellus intermedius *as a function of mean initial mg dry weight per individual. Error bars denote one standard deviation.*

converge as size increases. Regressions of growth versus time for each size class tested (Fig. 1) showed that all size classes grew significantly slower in the acidified environment than in the control stream except for the 4.31 ± 0.08 mg per individual size class (f statistic, $p < 0.05$). However, growth rate was very near zero for these sexually mature individuals in both the control and pH 4 streams.

Growth rate was dependent on temperature (Fig. 3) as well as pH depression. The growth rate was reduced for the pH 4 treatment (Fig. 3), and there was a seasonal component to temperature effects (Fig. 3). During periods of decreasing temperature from autumn to winter, the regression equation for the control stream was $y = 0.007 + 0.0009x$ ($R^2 = 0.85$, Fig. 3) (y = growth rate, x = temperature), while the equation for the pH 4 stream was $y = 0.003 + 0.0007x$ ($R^2 = 0.61$). The slopes of the lines were significantly different (f statistic, $p < 0.05$). During times of increasing temperature from late winter to spring, the regression equation for the control stream was $y = -0.024 + 0.002x$ ($R^2 = 0.90$, Fig. 3), while the regression for the pH 4 stream was $y = -0.009 + 0.0007x$ ($R^2 = 0.20$, Fig. 3—very few individuals survived at all so growth rate was obtained for only three samples). These highly significant seasonal differences may have simply reflected differences in the age of individ-

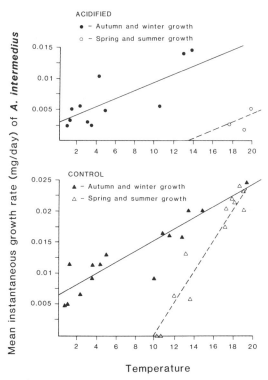

FIG. 3—*Regression of temperature with mean instantaneous growth rate of* Asellus intermedius.

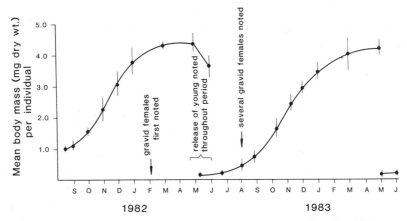

FIG. 4—*Growth of* Asellus intermedius *collected from streams near Paradise, Michigan, and maintained in recirculating laboratory streams at ambient temperatures and pH levels (6.7–7.2). Error bars are 95% confidence intervals.*

uals in the population. Small, rapidly growing individuals were present in the autumn, while larger, almost sexually mature individuals were present in the spring (Fig. 4).

Growth rates of *Asellus* collected from the streams near Paradise, Michigan and maintained under control conditions in the laboratory are depicted in Fig. 4. It is possible that rapidly growing and molting individuals would be more susceptible than would slow growing and molting individuals [8]. The least susceptible individuals of *Asellus* were the slow growing adults (Fig. 1). However, the slow growing, newly hatched 0.20 mg ± 0.01 individuals (Fig. 4) were the most susceptible (Fig. 1). Thus, size alone appears to determine the degree of susceptibility of *Asellus*.

Lepidostoma liba

As was true of *Asellus*, mortality was significantly greater in the pH 4 stream than in the control stream for all size classes of *Lepidostoma* tested for each interval tested (Fig. 5, t test, $p < 0.05$). The one exception was for the 1.00 ± 0.06 mg per individual size class because of unexplained mortality in the control stream after 5 weeks (Fig. 5). The trend of significantly decreased survival with decreased size described for *Asellus* was also characteristic of *Lepidostoma* (f statistic, $p < 0.05$; Fig. 5). Complete mortality occurred in less than 2 weeks in the pH 4 stream for the 0.05 ± 0.005-mg individuals with length of time of at least some surviving specimens slowly increasing as body size increased (Fig. 5). As size increased to 0.64 ± 0.05 mg or greater, more than half the individuals survived the complete 8 weeks exposure and differences between the control stream and the pH 4 stream decreased (Fig. 5). The trend of increasing mortality with time

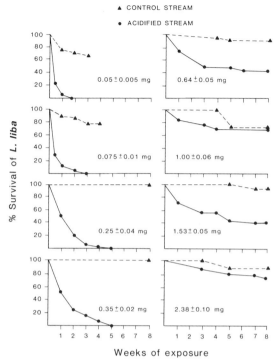

FIG. 5—*Percent survival of* Lepidostoma liba *by size class (mean initial individual mg dry weight ± one standard deviation).*

for each size class tested when compared to the control was statistically significant over all classes (f statistic $p < 0.01$).

The growth of *Lepidostoma* collected from the streams near Paradise, Michigan and maintained in the control laboratory stream at ambient temperatures was related to temperature (Fig. 6). The sigmoid growth response in the autumn was interrupted by a slow period during colder winter months with rapid growth resuming in the spring as temperatures increased (Fig. 6).

Susceptibility of the larvae appeared to be influenced by rate of growth as well as size. The 1.00 ± 0.06 mg per individual size class tested during the period of slow growth in winter (Fig. 6) was much less susceptible to the pH 4 treatments than was either the smaller 0.66 ± 0.05 individuals tested during rapid growth in the autumn or the larger 1.53 ± 0.05 individuals tested during rapid growth in spring (Fig. 5). Likewise, the 2.38 ± 0.10-mg individuals were only slightly susceptible to pH 4 depression (Fig. 5), and these individuals were growing very slowly prior to pupation (Fig. 6).

Growth rate of *Lepidostoma* was related to both temperature and acidity with marked seasonal differences. During periods of decreasing temperatures from

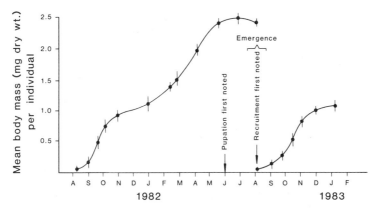

FIG. 6—*Growth of* Lepidostoma liba *collected from streams near Paradise, Michigan, and maintained in recirculating laboratory streams at ambient temperatures and pH levels (6.7–7.2). Error bars are 95% confidence intervals.*

autumn to winter, the regression equation for the control stream was $y = -0.0009 + 0.003x$ ($R^2 = 0.59$, Fig. 7) (y = growth rate, x = temperature), while the regression for the pH 4 stream was $y = 0.004 + 0.0004x$ ($R^2 = 0.17$, Fig. 7). The slopes of the lines were significantly different (f statistic, $p < 0.05$), indicating decreased growth in the pH 4 stream. During increasing temperatures from late winter to spring, the regression for the control stream was $y = 0.009 - 0.0005x$ ($R^2 = 0.58$, Fig. 7), and the regression for the pH 4 stream was $y = 0.008 - 0.004x$ ($R^2 = 0.16$). Thus, temperature affected growth minimally during spring. Correlations of temperature with growth were poor for this species in either season, and differences in growth were primarily the result of acidity and the weight of the individuals tested (Fig. 8). No specimens survived in the pH 4 stream for 28 days for the smallest size classes (Fig. 5). Growth after 28 days was reduced significantly for surviving specimens in the pH 4 stream for the 0.25 ± 0.04 and 0.35 ± 0.002-mg dry weight per individual groups tested compared to the control stream (Fig. 8). For the 0.64 ± 0.05 and larger groups, growth of surviving specimens in the pH 4 stream was only slightly reduced compared to growth of individuals in the control stream (Fig. 8).

Physella heterostropha

Mortality of *Physella* was correlated with shell size (Fig. 9). Small individuals with a shell size of 4 mm were eliminated from the pH 4 stream by Day 14. Individuals of intermediate size (7 mm) also had significantly reduced survival rates (Fig. 9). However, mature specimens (12 mm) were able to survive as well in the pH 4 stream as in the control stream (Fig. 9).

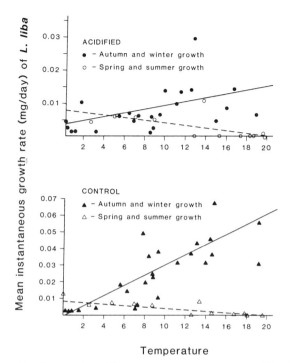

FIG. 7—*Regression of temperature with mean instantaneous growth rate of* Lepidostoma liba.

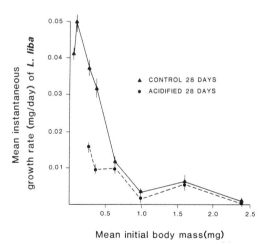

FIG. 8—*Mean instantaneous growth rate (mg/day) of* Lepidostoma liba *as a function of mean initial mg dry weight per individual. Error bars denote one standard deviation.*

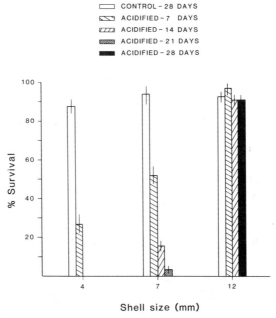

FIG. 9—*Survival of* Physella heterostropha *as a function of shell length. Survival was zero by day 14 for 4 mm and by day 28 for 7 mm specimens. Error bars denote one standard deviation.*

Physella adults deposited many egg masses in the control stream. Egg masses deposited in the pH 4 stream were rare. Thus, 14 egg masses (248 ova) from the control stream were moved from one locality in the control stream to another while 8 masses (134 ova) were moved from the control stream to the pH 4 stream. No ova survived in the pH 4 stream, while 84% of the ova in the control stream survived until hatching.

Discussion

Data from this study indicated that susceptibility of invertebrates to acidification was highly dependent on both size and growth rate. The smallest individuals of all three species were rapidly eliminated by exposure to pH 4. As size increased, survivorship increased. More rapid rates of growth for larger individuals during periods of increasing temperatures for at least *Lepidostoma* also appeared to decrease survivorship. Such size- and growth rate-dependent responses to the stress of acidification indicated that a program designed to detect sensitivity of invertebrates to an environmental stress should include testing of several life stages with special emphasis on smaller size classes. The complete elimination of the smallest size classes for the three species tested indicated that populations of these species would be eliminated in soft-water streams subjected to a sustained pH depression from 7 to 4. A very different prediction would have emerged if only the larger size classes had been tested. A pulse of acidification from a storm or from snowmelt also would have differential effects on the three populations depending on the size classes exposed and their rates of growth.

Acknowledgments

This work was supported by Grant A-120-MICH, U.S. Department of Interior, Office of Water Policy, as authorized by the Water Research and Development Act of 1978 (P.L. 95-467). We thank the Institute of Water Research at Michigan State University for use of their facilities to house the streams.

References

[1] Bell, H., *Canadian Entomologist*, Vol. 102, 1970, pp. 636-639.
[2] Bell, H. L., *Water Research*, Vol. 5, 1971, pp. 313-319.
[3] Bell, H. L., and Nebeker, A. V., *Journal of the Kansas Entomological Society*, Vol. 42, No. 2, 1969, pp. 230-236.
[4] Burton, T. M., Stanford, R. M., and Allan, J. W. in *Acid Precipitation: Effect on Ecological Systems*, F. M. D'Itri, Ed., Ann Arbor Science Publishers, Ann Arbor, 1982, pp. 209-235.
[5] Haines, T. A., *Transactions of the American Fisheries Society*, Vol. 110, 1981, pp. 669-707.
[6] Hall, R. J., Likens, G. E., Fiance, S. B., and Hendrey, G. R., *Ecology*, Vol. 61, No. 4, 1980, pp. 976-989.
[7] Herricks, E. E. and Cairns, J., Jr., *Revista de Biologia*, Vol. 10, No. 1, 1974, pp. 1-11.
[8] Malley, D. F., *Canadian Journal of Fisheries and Aquatic Sciences*, Vol. 37, 1980, pp. 364-372.
[9] Raddum, G. E., and Steigen, A. L. in *Effects of Acidic Precipitation on Benthos*, R. Singer, Ed., North American Benthological Society, Illinois Environmental Protection Agency, Springfield, IL, 1981, pp. 97-101.
[10] Pratt, J. M. and Hall, R. J. in *Effects of Acidic Precipitation on Benthos*, R. Singer, Ed., North American Benthological Society, Illinois Environmental Protection Agency, Springfield, IL, 1981, pp. 75-77.
[11] Fiance, S. B., *Oikos*, Vol. 31, No. 3, 1978, pp. 332-339.
[12] *An Introduction to the Aquatic Insects of North America*, Merritt, R. W. and Cummins, K. W., Eds., Kendall/Hunt Publishing Co., Dubuque, 1978, p. 441.
[13] Wiggins, G. B., *Larvae of the North American Caddisfly Genera*, University of Toronto Press, Toronto, 1977, p. 401.
[14] Ross, H. H., *Bulletin of the Illinois Natural History Survey*, Vol. 23, No. 1, 1944, pp. 1-326.
[15] Pennak, R. W., *Fresh-Water Invertebrates of the United States*, John Wiley & Sons, New York, 1978, p. 803.
[16] Williams, W. D., "Freshwater Isopods (Asellidae) of North America," Second Printing, Water Pollution Control Research Series Technical Report 18050 ELDO5/72, U.S. Environmental Protection Agency, Cincinnati, 1976, p. 45.
[17] Harman, W. N. and Berg, C. O., *Search Agriculture*, Cornell University Agricultural Experiment Station, Vol. 1, No. 4, 1971, p. 67.
[18] Burch, J. B., "Freshwater Snails (Mollusca: Gastropoda) of North America," Environmental Monitoring and Support Laboratory Report EPA-600/3-82-026, U.S. Environmental Protection Agency, Cincinnati, 1982.

[19] Burton, T. M., Stanford, R. M., and Allan, J. W., *Canadian Journal of Fisheries and Aquatic Sciences,* Vol. 42, No. 4, 1985, pp. 669–675.
[20] Cummins, K. W., Peterson, R. C., Howard, F. O., Wuycheck, J. C., and Holt, V. I., *Ecology,* Vol. 54, No. 2, 1973, pp. 336–345.
[21] Grafius, E. J. and Anderson, N. H., *Ecology,* Vol. 61, No. 4, 1980, pp. 808–816.
[22] Gill, J. L., *Design and Analysis of Experiments in the Animal and Medical Sciences,* Iowa State University Press, Ames, 1978, p. 409.

Todd E. Perry,[1] Curtis D. Pollman,[2] and Patrick L. Brezonik[1]

Buffering Capacity of Soft-Water Lake Sediments in Florida

REFERENCE: Perry, T. E., Pollman, C. D., and Brezonik, P. L., **"Buffering Capacity of Soft-Water Lake Sediments in Florida,"** *Impact of Acid Rain and Deposition on Aquatic Biological Systems, ASTM STP 928,* B. G. Isom, S. D. Dennis, and J. M. Bates, Eds., American Society for Testing and Materials, Philadelphia, 1986, pp. 67–83.

ABSTRACT: Fifty sediment cores from 13 soft-water lakes in north- and south-central Florida were collected to examine the role of surficial sediments in buffering lake water against changes in pH induced by additions of sulfuric acid (H_2SO_4). Intact cores were collected in duplicate from nearshore and pelagic locations to stimulate in situ lake sediment-water conditions. In addition, batch titrations of sediments treated with chloroform were performed in continuously shaken sediment-water slurries to evaluate ion exchange mechanisms and total mineralogical potential for acid neutralization.

Results from batch bottle experiments showed higher pH values in sediment-water slurries than corresponding lake water pH, suggesting that chemical processes in the sediments tend to counter inputs of acidity to the water column. The acid-neutralizing capacity (ANC) was moderately related to organic content; consequently, ANC was more pronounced in pelagic sediments, which typically accrete organic matter, than littoral sediments, which are more subject to wind-induced scour and resuspension and characteristically are impoverished in organic matter. Calculated areal buffering capacities from the batch studies ranged from 5.5 to 149 meq/m^2/cm, indicating that sufficient buffering is available in 1 to 2 cm of sediment to neutralize annual proton loadings via ion exchange alone.

Significant neutralization of acid inputs also was observed in intact cores, and analysis of changes in cation chemistry in the overlying water of the cores suggests that microbially mediated sulfate reduction exceeded ion exchange as the principal alkalinity-producing mechanism. High rates of sulfate removal from the experimental cores support this conclusion; sulfate losses ranged from 32 to 94% in pelagic cores and 39 to 91% in littoral cores. Cores that received no acid showed losses of sulfate ranging from 55 to nearly 100% with concomitant increases in pH in the overlying water. Neutralization of added acid directly attributable to the sediments ranged from 31.8 to 101 meq/m^2 over the 4-month duration of the study. Neutralization was independent of particle size distribution or sediment organic content.

KEY WORDS: acidic deposition (acid rain), sediments, acid-neutralizing capacity (ANC), pH, buffering, sulfate reduction

---

[1] Graduate research assistant and professor, respectively, Department of Civil and Mineral Engineering, University of Minnesota, Minneapolis, MN 55455.

[2] Head, Environmental Chemistry Department, Water Resources Division, Environmental Science and Engineering, Inc., Gainesville, FL 32602.

Within Florida, lakes exhibiting low pH and alkalinity are associated with the highlands and ridge regions of the state. Most prominent are the Western Highlands of the panhandle and the Central Ridge, which includes the Trail Ridge and Highlands or Lake Wales Ridge lake districts in north- and south-central Florida, respectively. Lakes in these regions may be susceptible to acidification from atmospheric inputs for three reasons. First, soft-water lakes have low alkalinity. Thus, apparent in-lake hydrogen ion (H^+) neutralization capabilities are low. Second, watershed soils are extremely sandy and characteristically lack significant buffering capabilities. Third, many of the lakes have no surface inflows or outflows and receive their water primarily from atmospheric inputs directly to the lake and from seepage from the shallow (water table) aquifer. In general, seepage flows are thought to be small compared to direct precipitation, and the water and mineral budgets of seepage lakes are dominated by direct precipitation. Consequently, surrounding terrestrial areas may not be important in neutralizing lacustrine inputs of acidity from the atmosphere.

To determine the effects of acidic deposition on lake ecosystems, quantifying dose/response relationships is critical [1]. Most existing models to predict the degree of acidification do not consider the inherent buffering or neutralizing capacity of sediments afforded to the overlying water. Recent studies on McCloud Lake, Florida, suggest, however, that bottom sediments may buffer the overlying water significantly with respect to additions of both strong acid and base [2]. Sediments may neutralize acid (or generate alkalinity) by four potential mechanisms:

1. Ion exchange; that is, replacement of base cations on sedimentary clays and organic matter by H^+ from the water.
2. Mineral dissolution involving clays (aluminosilicates), oxides, and carbonates.
3. Sulfate adsorption onto oxide surfaces, which can occur by reversible (ion exchange) or irreversible mechanisms.
4. Biological reduction of sulfate and subsequent loss of hydrogen sulfide (H_2S) by volatilization or precipitation as heavy metal sulfides [for example, iron sulfide (FeS)].

The first and fourth mechanisms are likely to be most important in the sediments of soft-water lakes in Florida. Although littoral sediments of these lakes are primarily unreactive (quartz) sands, as are the soils surrounding the lakes, even the sandiest soils have a small clay content. For example, typic quartzipsamment surficial soils in the region have cation exchange capacities (CECs) of 1 meq/100 g. Littoral sediments, which appear similar to surrounding terrestrial soils, presumably also have some clay content and a low CEC. Organic-rich sediments in the central basins of most lakes likely have higher CEC values; soil organic matter has CEC levels of 120 to 300 meq/100 g.

Sulfate reduction was found to be an important mechanism for neutralization of acid in Lake 223 in the Experimental Lakes Area (ELA) of western Ontario.

This lake was the subject of an artificial acidification experiment for more than 5 years [3]. The investigators found that they consistently had to add more sulfuric acid (H_2SO_4) than predicted based on lake alkalinity and volume to achieve their target pH values. During the acidification experiment, about 50% of the sulfate was lost from the lake, primarily by reduction to sulfide [4]. Lake 223 stratifies and develops an anoxic hypolimnion during summer, and some sulfate reduction occurred there. Most of the reduction and thus most of the acid neutralization occurred in anoxic sediments, however, as sulfate diffused from the overlying water into the sediments.

Recent studies in northern Florida [2,5] have demonstrated that sedimentary reduction of sulfate is an important sink for sulfate and acidity in a small, unstratified acidic lake (McCloud Lake) near Melrose. The lake's sediments do not adsorb significant quantities of sulfate, implying that this mechanism is not important in neutralizing acidity [2].

Mineral dissolution processes are unlikely to contribute significantly to acid neutralization in Florida lakes because primary minerals (for example, feldspars) are absent from Florida soils and lake sediments, and secondary minerals (aluminosilicate clays) are not abundant in surficial soils or in lake sediments [6,7]. The acidic nature of the soils within the watershed of most soft-water lakes precludes the existence of significant quantities of carbonate minerals [for example, calcium carbonate ($CaCO_3$)]. Minor quantities of $CaCO_3$ may enter the lakes by wet and dry deposition of wind-blown calcareous soil particles originating in coastal and southern Florida. Calcareous shells (for example, freshwater clams and mussels) are absent or rare in the soft-water lakes of northern Florida; the combination of low calcium levels and low pH prevents these organisms from synthesizing calcareous shells.

Preliminary studies on sediment buffering mechanisms in Lake McCloud [8] found that neutralization of acid added to sediments was rapid (equilibrium in less than 24 h), suggesting that rapid ion-exchange processes were responsible for the observed acid neutralization. These preliminary studies, however, were conducted under well-mixed conditions not conducive for alkalinity generation by sulfate reduction. The possibility of further buffering by slow-acting mineral dissolution processes (at time scales longer than those of the experiments) also may be important.

For highly acidic lakes, where the bicarbonate buffering system may be exhausted (that is, at pH <5.0), dissolution of aluminum (Al) and iron (Fe) minerals in sedimentary deposits may constitute an important buffering mechanism against further decreases in pH. This type of mechanism, however, results in the release of Al to the water column, the effect of which has received much attention because of Al toxicity to fish [9].

Objectives

Because of the need to better quantify the response of potentially sensitive Florida lakes to acid deposition, this study was designed to evaluate the importance

70 IMPACT OF ACID RAIN

of sediment ion exchange and sulfate reduction mechanisms in neutralizing acidity in soft-water Florida lakes. In addition, a corollary objective was to characterize sediment types and determine associated relationships with buffering mechanisms. In brief, the study seeks to identify mechanisms of pH buffering provided by lake sediments and evaluate their importance in neutralizing acid inputs to lakes.

Methodology

General

Duplicate sediment samples were obtained from littoral and pelagic sites in 13 lakes of the Trail Ridge and Highland Ridge regions of north-central and south-central Florida (Fig. 1) by a Livingstone-type piston coring device with 4.1-cm-inside-diameter by 30-cm acrylic coring tubes. The lakes ranged in pH from

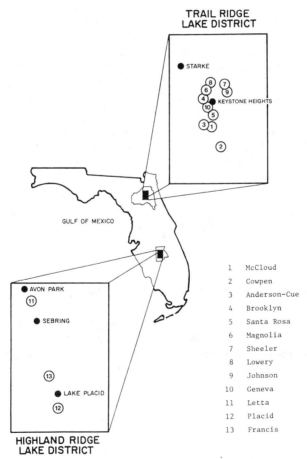

FIG. 1—*Locations of the two lake groups sampled in the survey.*

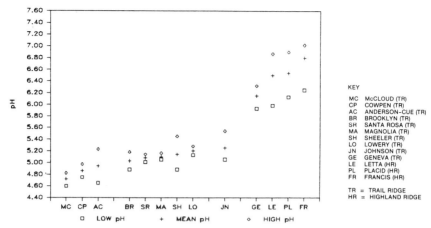

FIG. 2—*Mean and range of pH values in lake selected for sediment study.* NOTE: *values from Brezonik et al.* [11].

approximately 4.60 to 7.00 (Fig. 2), and all lakes had soft water with low alkalinity (generally <100 μeq/L, all lakes <240 μeq/L [10]). A parafilm-covered rubber stopper was used to seal the coring tube after removal from the lake bottom. Approximately 10 cm of sediment and 20 cm of overlying water depth were collected. Laboratory experiments with these sediments were conducted on intact cores and on extruded sediments in well-mixed bottles. Intact sediment cores were used to simulate in situ lake sediment-water conditions and evaluate the extent and mechanisms of acid neutralization in the lakes. More detailed bottle studies were done with continuously shaken sediment-water slurries to quantify ion exchange processes and evaluate the total mineralogical potential for acid neutralization.

Batch Bottle Experiments

Sediment samples used in well-mixed (oxygenated) batch experiments were obtained by decanting the overlying water in the field and extruding the top 10 cm of sediment, which was placed into a Whirl-pak bag. In the laboratory, weighed aliquots (200 g wet weight) of homogenized sediment from each core were placed in separate 1-L polyethylene bottles (25 samples total) to which 10 mL of 0.75% chloroform ($CHCl_3$) to inhibit microbial growth and 500 mL of unfiltered lake water were added. In addition, two batch bottles without sediment were run concurrently to serve as blanks and to examine possible losses of H^+ and SO_4^{-2} to the polyethylene bottles. A mass balance analysis of the reactor bottles indicated that H^+ and SO_4^{-2} were conservative in the absence of sediment. All bottles were placed on a rotary mixer for 1 week prior to removal of an aliquot of water for initial chemical analysis. Seven sequential acid additions of 0.135 N H_2SO_4 (1.0 to 2.5 mL each) were made at approximately 1-week intervals. Acid addition was terminated when the water pH remained below 4.0. Upon completion of the batch bottle titrations, the sediment-water slurry was transferred to a 500-mL beaker, dried at 105°C, and combusted at 550°C for 4 h for organic matter (volatile solids) determinations. Particle size determinations for sand (2.0 to 0.005 mm), silt (0.05 to 0.002 mm), and clay (<0.002 mm) were conducted by the soil hydrometer method [12,13]. Homogenized sediment was treated with 30% hydrogen peroxide (H_2O_2) to oxidize organic matter prior to particle size determination.

Intact Core Experiments

Fifty cores were set up in the laboratory for the 13 lakes—two sites per lake (except Lake Sheeler with only a littoral site) and two cores per site. One duplicate core served as a control (no acid additions); the other received periodic additions of 0.135 N H_2SO_4. Seven volumetric additions of acid ranging from 48 to 327 µL were made over 4 months; each addition was sufficient to lower the overlying water pH to 4.0. The pH of the overlying water was monitored biweekly by transferring 2.5 mL of column water to a distilled water washed Autoanalyzer vial and determining the pH with an Orion Model 811 Digital pH meter. Subsequent acid additions were made by calculating the amount of acid needed to lower the overlying water to the target level of pH 4.0, based on the present pH level and the volume of the overlying water. Previous studies in this laboratory indicated a gradual rise in overlying water pH in sediment-water systems, and a steady-state condition is reached within 1 week.

Following a 4-week acclimation period and prior to acid additions, a 45-mL aliquot from the overlying water of each intact core was removed, filtered, and analyzed for cation and anion parameters. These samples served in establishing preacidification levels for the various chemical parameters examined. Following the seventh and final acid addition, aliquots of overlying water were removed from each intact core to quantify postacidification cation and anion concentrations.

Gelman-type A/E glass fiber filters, washed with 3% American Chemical Society (ACS) reagent-grade nitric acid (HNO_3), were used to filter samples for analysis of dissolved cations. Flame atomic absorption spectrophotometry (Varian Model 175) was used to determine calcium (Ca^{+2}), magnesium (Mg^{+2}), potassium (K^+), and sodium (Na^+) concentrations. Nuclepore 0.45-µm polycarbonate filters were used to filter aliquots for anion samples. Automated AutoAnalyzer methods were used to determine SO_4^{-2} (methylthymol blue) and Cl^- (ferric thiocyanate).

Results

Physical and Chemical Characteristics of Sediments

Acid neutralization can occur in the absence of biological-mediated reactions by mineral dissolution and cation exchange. Organic and clay particles are negatively charged in the pH range of natural waters and sediments. The negative charges are balanced by exchangeable cations held to the particles by electrostatic forces. Increases in H^+ in solution induce displacement of exchangeable bases by protons, causing an increase in solution concentration of cations and a rise in pH.

Organic matter (OM) content for each sediment type is shown in Table 1. Sediment OM content tends to be either quite low (<6%) or rather high (35 to 85%). Most sediments were sandy and in the low OM range. Of the high OM sediments, all but one came from a pelagic site. This reflects the fact that organic material tends to accumulate in the deepest part of a lake basin. Wind and wave scouring tends to resuspend and move material from the littoral shallows, with net sedimentation occurring in more quiescent deep water. Localized pockets of high OM sediment can occur near shore in some lakes. In one case among the present samples (Lake Geneva), the littoral sampling site was adjacent to a cattail-sedge community, and a large quantity of plant detritus was noted in the sediment.

Particle size analyses (Table 1) indicate that most sediments are dominated by the sand fraction. Of the sediment mineral content, sand makes up between 88.5 and 98.8% by weight. The silt fraction accounted for 10 to 20% and clays accounted for only 0.1 to 6% by weight. Although clays were the smallest fraction of dry mass in all sediments, they may be important in terms of potential buffering because of larger surface areas and the presence of negatively charged functional groups for cation exchange. However, aluminosilicate clay in Florida is predominantly kaolinite [$Al_2Si_2O_5(OH)_4$], which has a low CEC (1 to 10 meq/100 g) [14]. Inorganic ion exchange involving hydrogen ions thus should be relatively low in the sediments, especially in those with low clay content. In contrast, natural OM contains carboxyl, phenolic, and hydroxyl functional groups, and soil/sedimentary OM typically exhibits exchange capacities of 50 to 300 meq/100 g between pH 5.5 and 4.5 [15]. Sediment OM thus may contribute more to acid neutralization by cation exchange in these sediments than do kaolinitic clay minerals.

Batch Bottle Experiments

Following 1 week of continuous shaking prior to acid addition, all sediment-water slurries had higher pH values than the corresponding ambient lake water. These results indicate that some process in the sediments is acting to neutralize H^+ in the lake waters. The sediment-water slurries showed varying degrees of pH depression with each sequential addition of acid. Littoral sediments generally had greater decreases in pH than did pelagic sediments (Fig. 3).

Acid-neutralizing capacities (ANC) were calculated graphically for each sediment on a 100-g dry-weight basis for the pH ranges 5.5 to 5.0, 5.0 to 4.5, and 4.5 to 4.0 from the amounts of acid added and the equilibrium pH following acid addition (Table 2). ANC values were summed over the pH range 5.5 to 4.5 as a measure of the neutralization possible through abiotic mechanisms in the pH region of greatest interest. [No uncolored soft-water lakes in Florida are known

TABLE 1—*Particle size distribution and organic content of littoral (L) and pelagic (P) sediments from 13 Florida lakes.*

Lake	Sediment Type[a]	Organic Matter, %	Sand, %	Silt, %	Clay, %
Anderson-Cue	L	4.83	93.8	4.5	1.6
	P	73.85	88.8	9.0	2.2
Brooklyn	L	0.13	96.7	1.6	1.7
	P	3.39	94.7	2.0	3.3
Cowpen	L	0.63	96.7	2.5	0.8
	P	0.76	96.3	2.9	0.8
Francis	L	5.63	98.3	...[b]	1.7
	P	1.53	98.8	0.4	0.8
Geneva	L	36.76	89.7	7.8	2.5
	P	1.63	97.5	1.3	1.2
Johnson	L	1.21	98.3	1.4	0.3
	P	3.75	96.7	2.8	0.5
Letta	L	1.21	97.5	0.8	1.7
	P	35.38	88.5	6.5	5.0
Lowery	L	0.06	98.8	1.2	...[c]
	P	1.03	95.8	2.0	2.2
Magnolia	L	0.73	95.8	2.0	2.2
	P	0.40	95.8	2.5	1.7
McCloud	L	5.00	...[b]	...[b]	...[b]
	P	85.51	...[b]	...[b]	...[b]
Placid	L	0.39	96.7	3.0	0.3
	P	1.02	98.3	0.9	0.8
Santa Rosa	L	0.45	97.5	1.7	0.8
	P	0.88	96.7	2.8	0.5
Sheeler	L	0.38	97.5	2.5	...[c]

[a] L = littoral; P = pelagic.
[b] Not determined.
[c] Negligible.

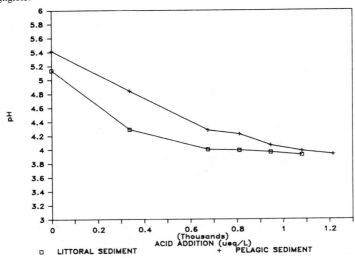

FIG. 3—*Sediments-water pH versus amount of strong acid (H_2SO_4) added in batch experiment for Anderson-Cue Lake.*

TABLE 2—*Ion-exchange acid-neutralizing capacities over various pH ranges from batch experiments. ANC values are expressed both on a mass [meq/100 g (dry weight)] and an areal basis (meq/m²/cm^b).*

Lake	Sediment Type[a]	ANC 6.0 to 5.5	ANC 5.5 to 5.0	ANC 5.0 to 4.5	ANC 4.5 to 4.0	SUM ANC 5.5 to 4.5	Estimated Bulk Density, g/cm³	ANC, meq/m²/cm[b]
Anderson-Cue	L	ND	0.03	0.13	0.32	0.16	1.71	19.3
	P	ND	1.75	2.40	3.65	4.15	1.01	27.3
Brooklyn	L	0.02	0.07	0.09	0.18	0.16	1.96	24.7
	P	0.03	0.15	0.25	0.39	0.40	1.53	35.1
Cowpen	L	0.02	0.08	0.10	0.27	0.18	1.97	28.3
	P	0.03	0.08	0.13	0.35	0.21	1.94	32.1
Francis	L	ND	ND	0.17	0.61	0.17	1.46	13.2
	P	0.16	0.20	0.21	0.21	0.41	1.85	57.0
Geneva	L	0.02	0.06	0.16	0.25	0.22	1.10	5.5
	P	ND	0.13	0.23	0.42	0.36	1.76	44.6
Johnson	L	0.02	0.05	0.08	0.16	0.13	1.94	19.9
	P	ND	0.11	0.18	0.27	0.29	1.76	36.7
Letta	L	0.02	0.10	0.14	0.13	0.24	1.94	36.6
	P	0.90	1.80	5.92	6.18	7.72	1.08	149.4
Lowery	L	ND	0.05	0.08	0.15	0.13	2.07	22.4
	P	0.02	0.10	0.17	0.22	0.27	1.82	35.9
Magnolia	L	ND	0.02	0.05	0.07	0.07	1.86	9.8
	P	ND	0.04	0.05	0.08	0.09	1.87	12.7
McCloud	L	ND	0.02	0.20	0.46	0.22	1.72	26.9
	P	0.54	2.15	3.69	5.31	5.84	1.02	20.2
Placid	L	0.08	0.09	0.09	0.09	0.18	1.93	27.0
	P	0.11	0.15	0.20	0.31	0.35	1.83	47.3
Santa Rosa	L	ND	0.05	0.06	0.10	0.11	1.94	16.7
	P	ND	0.06	0.09	0.24	0.15	1.84	20.3
Sheeler	L	ND	0.03	0.03	0.03	0.06	1.97	9.4

NOTE: ND = Not determined because initial pH was below 5.5.
[a] L = littoral, P = pelagic.
[b] meq/m²/cm = milliequivalents per square metre per centimetre.

to have pH values <4.5, and game fish such as largemouth bass occur in lakes with pH 4.5 or less.] Most of the sediments had ANC values <0.41 meq/100 g over this pH range, and only three sediments had ANC values ≥4.15 meq/100 g.

A moderate relationship was found between ANC and percent OM for the sediment samples: ANC = 0.063 (percent OM) + 0.168; r^2 = 0.631, N = 25. The relationship between ANC and percent clay is less well defined: ANC = 1.021 (percent clay) − 0.761; r^2 = 0.463, N = 24. OM in soft-water Florida lake sediments thus seems to have a greater influence than clays on acid neutralization under experimental conditions that minimize effects of biological activity.

Based on changes in base cation concentrations following acid additions, divalent cations are much more important than monovalent cations in the ion exchange neutralization of H^+ (Fig. 4). The pattern displayed for selected lakes in Fig. 4 is valid for all the lake sediments; that is, $Ca^{+2} > Mg^{+2} >> Na^+ > K^+$. Increases in Ca^{+2} and Mg^{+2} in solution together account for more than 70% of the ANC in the batch experiments.

Intact Core Experiments

Intact cores of sediment and overlying water simulate the physical conditions in lakes more closely than batch studies with well-mixed sediment-water slurries and allow for observation of both biotic and abiotic processes. Sediment-water interactions involved in acid neutralization may be transport-limited (that is, limited by rates of diffusion of ions into and out of the sediment).

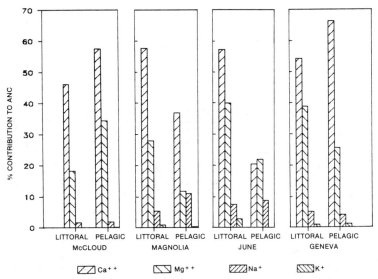

FIG. 4—*Percent contributions of major cations to ion-exchange-related ANC in littoral and pelagic sediments from representative lakes.*

Measurements of pH and major ion concentrations in the overlying water were made after a 4-week equilibration period (prior to the first additions of acid) and after the last addition of H_2SO_4, about 4 months later. Results are summarized in Table 3.

Initial concentrations in the overlying water of equilibrated cores were similar to values observed by Hendry [10] in the lakes in 1978–1979. Changes in overlying water chemistry after the last acid addition can be summarized as follows: (1) cation concentrations increased only slightly from beginning to end; (2) sulfate levels were much lower than predicted based on initial values plus H_2SO_4 additions (Fig. 5); (3) chloride concentrations were constant from beginning to end; and (4) measured pH was much higher than predicted based on the quantity of acid added (Fig. 6). Substantial neutralization of acid occurred in the cores during the experiment. For example, the pH of overlying water in the Anderson-Cue littoral core never dropped below 4.8 (measured 1 to 2 weeks after acid addition), even though each addition was calculated to lower the water pH to 4.0, and the seven additions were sufficient to lower the water pH to 3.1. The potential mechanisms

TABLE 3—*Summary of results of intact core titrations with 0.135N H_2SO_4. ANC values represent amount of acid consumed on an areal basis over the 4-month duration of the study and are corrected for ANC initially present in the overlying water.*

Lake	Sediment Type[a]	Initial pH	Final pH	Initial SO_4^{-2}, meq/L	Final SO_4^{-2}, meq/L	SO_4^{-2} Added, µeq	ANC, meq/m²
Anderson-Cue	L	5.23	5.06	0.04	0.15	100.6	58.4
	P	5.23	4.74	0.10	0.12	73.2	61.1
Brooklyn	L	5.07	4.20	0.06	0.27	94.2	61.2
	P	5.66	4.87	0.04	0.20	113.5	76.0
Cowpen	L	5.24	4.62	0.14	0.30	124.5	79.9
	P	5.28	6.46	0.11	0.14	100.8	75.3
Francis	L	6.69	5.03	0.13	0.36	152.3	88.5
	P	6.96	6.66	0.07	0.18	115.3	60.0
Geneva	L	6.56	6.13	0.10	0.10	147.8	101
	P	6.60	5.50	0.15	0.55	95.6	54.8
Johnson	L	5.96	4.37	0.06	0.22	90.9	31.8
	P	5.81	5.86	0.04	0.33	108.5	75.0
Letta	L	6.70	6.09	0.04	0.22	134.7	84.0
	P	7.06	6.68	0.14	0.35	136.9	77.3
Lowery	L	6.26	4.34	0.06	0.30	114.1	75.1
	P	6.04	4.33	0.04	0.21	113.0	68.1
Magnolia	L	4.98	4.85	0.04	0.29	61.8	41.8
	P	5.07	4.75	0.04	0.26	88.4	55.0
McCloud	L	4.66	4.47	0.06	0.14	63.9	42.6
	P	4.94	6.21	0.08	0.06	88.6	73.5
Placid	L	6.57	4.46	0.01	0.43	248.7	97.4
	P	7.14	4.67	0.02	0.62	252.6	79.4
Santa Rosa	L	4.74	4.32	0.02	0.26	109.2	56.4
	P	5.19	4.42	0.04	0.32	108.0	77.6
Sheeler	L	6.34	4.52	0.02	0.34	124.7	79.8

NOTE: ND = Not determined because initial pH was below 5.5.
[a] L = littoral, P = pelagic.

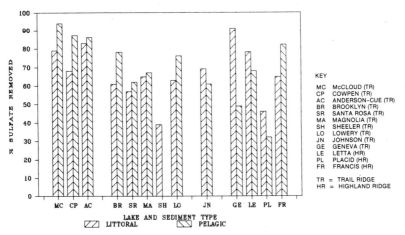

FIG. 5—*Percent loss of sulfate for overlying water in intact littoral and pelagic cores from the 13 lakes. Percent sulfate removed = (predicted − measured)/predicted × 100.*

for neutralization are ion exchange and sulfate reduction. Based on the findings that (1) cation concentrations changed only slightly during the experiment and (2) substantial losses of sulfate occurred (see Fig. 5), the principal mechanism apparently is biological sulfate reduction.

Sulfate reduction occurs under anoxic (strongly reducing) conditions; bacteria that use sulfate as a terminal electron acceptor under these conditions oxidize organic matter to obtain energy. The result is that a strong acid (H_2SO_4) is replaced by a weak acid (H_2S), which may be lost to the atmosphere by volatilization or (more likely) precipitated as FeS. If the ultimate source of Fe in lake sediments is terrestrial ferric oxyhydroxides, precipitation of FeS causes a net consumption of protons; that is, hydroxide released from iron oxyhydroxides reacts with H^+ released from H_2S. The net reaction entails generation of alkalinity, which can be demonstrated readily from charge balance considerations.

Figures 5 and 6 show that both trends occurred when H_2SO_4 was added to intact sediment-water cores from the 13 Florida lakes. The pH decrease of the overlying water (Fig. 6) was much less than that predicted from initial pH and alkalinity and the amount of added acid. Sulfate concentrations in the overlying water were much lower than values predicted from initial concentrations plus amounts of H_2SO_4 added. The percent sulfate removed ranged from 32 to 94% in pelagic sediments and from 39 to 91% in littoral sediments; thus, significant sulfate removal occurred in all cases. Control cores that received no acid also showed losses of sulfate ranging from 55 to nearly 100%, and the overlying water pH in control cores increased during the incubation. Ambient SO_4^{-2} concentrations in the controls ranged from about 2 to 12 mg/L. Disappearance of sulfate from the controls was expected; the same mechanisms account for losses in control cores (and whole lakes) as occurred in the acid-treated cores.

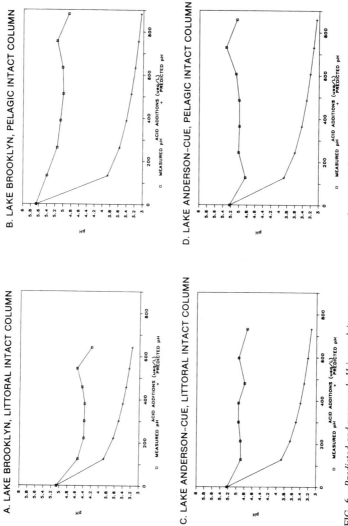

FIG. 6—*Predicted and measured pH in overlying water versus amount of strong acid to several intact cores. NOTE: littoral core from Anderson-Cue Lake exhibited greater buffering than pelagic core; reverse was true in most lakes.*

Because sulfate-reducing bacteria produce alkalinity and thus create a more basic environment, it is possible that sediment pH levels never decreased to the point where mineral dissolution and cation exchange could take place to a significant extent. In situ pore-water profiles in several soft-water Minnesota and Wisconsin lakes show a rapid increase in pH within the first several centimetres of sediment [5,16]. Several factors control the rate of sulfate reduction in sediments and thus affect the importance of this process in neutralizing acid inputs to lakes. The reduction process itself is biological and is affected by bacterial density, availability of carbon substrate (food supply), temperature, and redox status of the sediment. Sulfate reducers require strongly reducing conditions. The rate-controlling step in many lakes probably is mass transport, that is, diffusion of sulfate from oxygenated water into the anoxic zone of the sediments. The rate of diffusion can be predicted by Fick's law and is proportional to the concentration gradient between the water and anoxic sediment zone. As a first approximation, the gradient should be proportional to the sulfate in the lake water. Other factors being equal, sulfate reduction rates and thus acid neutralization rates should increase with increasing sulfate concentration. This may be an important example of homeostasis (that is, feedback control) in maintaining stable conditions in aquatic ecosystems.

Areal Buffering Capacities

The potential significance of ion exchange and biological sulfate reduction as acid neutralization mechanisms in lakes can be estimated by converting the laboratory results to areal capacities. Results for McCloud Lake will be used for the illustration. The littoral and pelagic sediments of this lake have estimated bulk densities of 1.72 and 1.00 g/cm^3, respectively, and corresponding ANCs for pH 5.5 to 4.5 from the batch experiments are 0.22 and 5.84 meq/100 g. The ion exchange neutralization capacity of the top centimetre of littoral sediment thus is about 27 meq/m^2, and the corresponding value for pelagic sediment is approximately 20 meq/m^2 (see Table 2). The annual (wet) deposition of protons to the surface of the lake is about 35 to 40 meq/m^2 [8,11]; thus, the top centimetre of both littoral and pelagic sediment has the potential to neutralize somewhat less than 1 year of proton deposition by ion exchange alone. These values should be considered in the context that the depth of exchange or interaction of surficial sediments with the water column typically extends several centimetres or more below the sediment-water interface; depending on the depth of exchange, total ion exchange neutralization capacity increases correspondingly. Calculated ion exchange neutralization capacities ranged from 5.5 meq/m^2/cm for Lake Geneva littoral sediments to 149 meq/m^2/cm for Lake Letta pelagic sediments, but most neutralization capacities were between 15 and 40 meq/m^2/cm. Because sediment bulk density increases with depth (because of compaction), the neutralizing capacity per centimetre of depth also increases with depth; this is especially the case for flocculent pelagic sediments.

Areal consumption rates of acid by the intact cores during the 4-month duration of the study ranged from 32 meq/m² for Lake Johnson littoral sediments to 101 meq/m² for Lake Geneva littoral sediments (Table 4). Results are corrected for the amount of buffering initially afforded by the water column itself. In contrast to the observed distribution of ion-exchange ANC, ANC associated with intact cores showed virtually no relationship with either sedimentary organic content or particle size distribution. Rather, it appears that the amount of ANC generated by intact sediments is related to the amount of SO_4^{-2} added. This is demonstrated in Fig. 7, which represents a modified Lineweaver-Burk approach to evaluating ANC generation as a SO_4^{-2} limited reaction. The modified approach reflects expressing SO_4^{-2} as the amount added or available to the sediment areally (that is, meq/m²) instead of using the traditional concentration expression (meq/L). Results suggest that sedimentary production of ANC follows Monod-type kinetics

$$\frac{1}{ANC} = \frac{1}{ANC_\infty} + \frac{K_s}{ANC_\infty} \cdot \frac{1}{SO_4^{-2}} \tag{1}$$

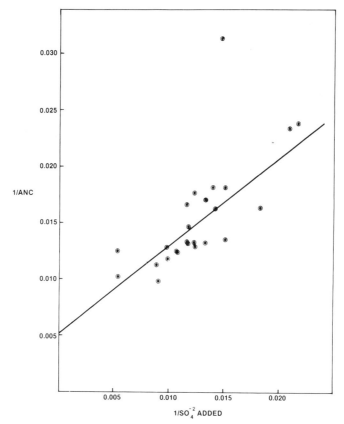

FIG. 7—*Lineweaver-Burk plot of ANC generation by intact cores as a function of SO_4^{-2} addition rates. Both parameters expressed as meq/m² over a time frame of 4 months.*

where ANC_∞ represents the asymptotic limit or maximum production rate of ANC and K_s is the half-saturation constant. Excluding the datum represented by Lake Johnson littoral sediments as an outlier yields a coefficient of determination of 0.72. The calculated value of ANC_∞ (over the 4-month duration of the study) was 194 meq/m^2, while K_s was determined to be 151 meq/m^2.

Although a similar response between SO_4^{-2} addition and ANC production would be expected from specific SO_4^{-2} adsorption following a Langmuir-type adsorption isotherm, several lines of evidence indicate that sulfate reduction is the principal mechanism. First, comparison of actual and expected SO_4^{-2} concentrations at the conclusion of the microbially inhibited batch studies indicated little or no loss of sulfate. Consequently, it is unlikely that SO_4^{-2} adsorption occurred during the intact core studies. This agrees with results of Baker [2], who determined that SO_4^{-2} adsorption by McCloud Lake sediments is unimportant. In addition, studies of sulfate adsorption by typic quartzipsamment soils (the dominant soil subgroup in the McCloud Lake watershed) show that sulfate adsorption is relatively minor, and the adsorption that does occur is nonspecific; the displaced anion is some ion other than OH^-, resulting in virtually no neutralizing effect on acidic inputs [17].

Conclusions

Studies on sediment cores from 13 softwater lakes within the Trail Ridge and Highland Ridge lake districts indicate that chemical and biological processes in the sediments tend to counter inputs of acidity to the water column. ANC attributable to ion-exchange mechanisms was strongly related to organic content; consequently, ion-exchange ANC was more pronounced in organic-rich pelagic sediments than in littoral sediments that are more subject to wind-induced scour and characteristically have low organic contents. Calculated areal buffering capacities due to ion-exchange ranged from 5.5 to 149 meq/m^2/cm, indicating sufficient buffering is available in 1 to 2 cm of sediment to neutralize annual proton loadings via ion exchange alone.

Analysis of changes in major ion chemistry in the overlying water of intact cores indicates that microbially mediated sulfate reduction exceeds ion exchange as the principal alkalinity-producing mechanism. This mechanism was supported by high rates of sulfate removal from both experimental and control cores. Neutralization of added acid attributable to the sediments ranged from 31.8 to 101 meq/m^2 over the 4-month duration of the study and appeared to follow Monod-type kinetics with SO_4^{-2} availability limiting ANC generation at low SO_4^{-2} concentrations. This study demonstrates that sediment buffering effects on lakewater pH may be appreciable; however, the ultimate fate of reduced sulfur and the relative importance of sulfate reduction in mitigating acidic deposition effects in seepage lakes still remains to be investigated.

References

[1] Gorham, E., "Acid Rain—An Overview," *American Chemical Society, Division of Environmental Chemistry,* Vol. 22, No. 1, 1982, pp. 23–25.

[2] Baker, L. A., "Mineral and Nutrient Cycles and Their Effect on the Proton Balance of a Softwater, Acidic Lake," Ph.D. thesis, University of Florida, 1983.

[3] Schindler, D. W. in *Atmospheric Sulfur Deposition: Environmental Impact and Health Effects,* D. S. Shriner, C. R. Richmond, and S. E. Lindberg, Eds., Ann Arbor Science, MI, 1980, pp. 453–462.

[4] Cook, R. W. and Schindler, D. W., "The Biogeochemistry of Sulfur in an Experimentally Acidified Lake," *Environmental Biogeochemistry,* Vol. 35, 1983, pp. 115–127.

[5] Baker, L. A. and Brezonik, P. L., "Fate of Sulfate in a Softwater Acidic Lake," in *Proceedings of the 6th International Symposium on Environmental Biogeochemistry,* Santa Fe, New Mexico, Van Nostrand Reinhold Publishers, New York, 1985, pp. 297–306.

[6] Carlisle, V. W., Caldwell, R. E., Sodek, F., Hammond, L. C., Calhoun, F. G., Granger, M. A., and Breland, H. L., "Characterization Data for Selected Florida Soils," Soil Science Research Report Number 78-1, University of Florida, Gainesville, FL. 1978.

[7] Carlisle, V. W., Hallmark, C. T., Sodek, F., Caldwell, R. E., Hammond, L. C., and Berkheiser, V. E., "Characterization Data for Selected Florida Soils," Soil Science Research Report Number 81-1, University of Florida, Gainesville, FL. 1981.

[8] Environmental Science and Engineering, Inc., Florida Acid Deposition Study, Phase III Report, Gainesville, FL, 1984.

[9] Wright, R. F., Conroy, N., Dickson, W. T., Harriman, R., Henriksen, A., and Schofield, C. L. in *Ecological Impact of Acid Precipitation,* D. Drablos and A. Tollan, Eds., SNSF Project, Sandefjord, Norway, 1980.

[10] Hendry, C. D., Jr., "Spatial and Temporal Variations in Bulk, Wet, and Dry Atmospheric Deposition of Acidity and Minerals Across Florida," Ph.D. thesis, University of Florida, 1983.

[11] Brezonik, P. L., Baker, L. A., Ogburn, R. W., Edgerton, E. S., and Crisman, T. L., "Ecological Effects of Acid Precipitation on Sensitive Softwater Lakes in Florida," EPA-NCSU Project No. APP-09007-02-1980, University of Florida, 1981.

[12] Day, P. R. in *Methods of Soil Analysis,* C. A. Black, Ed., American Society of Agronomists, Madison, Wisconsin, 1965, pp. 545–567.

[13] Grigal, D. F., "Note on the Hydrometer Method of Particle-Size Analysis," Minnesota Forest Resources Notes, No. 245, University of Minnesota, MN, 1973.

[14] Bohn, H. L., McNeal, B. L., and O'Connor, G. A., *Soil Chemistry,* Wiley-Interscience, New York, 1979.

[15] Helling, C. S., Chesters, G., and Corey, R. B., *Soil Science Society of America Proceedings,* Vol. 28, 1964, pp. 517–520.

[16] Baker, L. A., Perry, T. E., and Brezonik, P. L., unpublished data.

[17] Volk, R. G. and Pollman, C. D., "Sulfate Adsorption and Related Effects on Nutrient Dynamics in Poorly Buffered Soils," in preparation, 1986.

Eric L. Morgan,[1] Kenneth W. Eagleson,[2] Thomas P. Weaver,[3] and Billy G. Isom[4]

Linking Automated Biomonitoring to Remote Computer Platforms with Satellite Data Retrieval in Acidified Streams

REFERENCE: Morgan, E. L., Eagleson, K. W., Weaver, T. P., and Isom, B. G., "**Linking Automated Biomonitoring to Remote Computer Platforms with Satellite Data Retrieval in Acidified Streams,**" *Impact of Acid Rain and Deposition on Aquatic Biological Systems, ASTM STP 928,* B. G. Isom, S. D. Dennis, and J. M. Bates, Eds., American Society for Testing and Materials, Philadelphia, 1986, pp. 84–91.

ABSTRACT: Recent advances in site-specific detection of fish's physiological responses to continuous stream flows now complement surveillance programs utilizing remote automated water quality monitoring stations. Developments in automated biosensing provide methods for measuring in situ fish breathing rate changes to stream episodes and add a real-time biological monitoring dimension to remote water quality networks incorporating automated data collection platforms and satellite data retrieval options.

In meeting our objective of developing automated biosensing capabilities for remote monitoring, a series of field trials was designed to test various configurations of in situ fish-holding chambers, breathing rate detectors, and system interface to streamside water quality data collection platforms for satellite data retrieval.

Results were used to design groups of automated biosensing devices for detecting rainbow trout breathing rate responses and to implement eight units at each of two data collection, platform-equipped, water quality stations located along a stream subject to acid precipitation influences in the Southern Appalachian Mountains. Remote stations are being maintained for real-time data needs as a part of the 5-year Acid Precipitation Mitigation Program initiated in 1985 by the U.S. Fish and Wildlife Service to evaluate stream ecological responses to regulated liming operations.

KEY WORDS: acid deposition, automated biomonitoring, remote biosensing, remote computer satellite data retrieval, real-time data acquisition, water quality monitoring, biological monitoring, regional water quality network

---

[1] Center for the Management, Utilization, and Protection of Water Resources, Tennessee Technological University, Cookeville, TN 38505.

[2] Supervisor, Biological Services Unit, North Carolina Department of Natural Resources, Division of Environmental Management, Raleigh, NC 27611.

[3] Biologist, Office of Water Management, Tennessee Department of Health and Environment, Nashville, TN 37219.

[4] Program manager, Aquatic Research Laboratory, Tennessee Valley Authority, Decatur, AL 35602.

Methods for rapidly evaluating the biological integrity of multipurpose water resources are becoming increasingly important as water demands intensify. In this regard, water resource managers and those responsible for monitoring water quality have long been restrained in their ability to make timely decisions. This limitation in large part has been the result of imposed time and manpower requirements inherent in most biological monitoring efforts. A possible solution to help alleviate these restrictions and complement acid deposition monitoring programs would be to employ some sort of real-time automated biomonitoring device using a surrogate group of aquatic animals for detecting undesirable water quality changes. Until recently, such biomonitoring applications have been limited to specific research studies and a few industrial wastewater surveillance programs [1].

Presently, monitoring programs that include real-time transmission of key physical parameters from remote stations in a basin generate continuous information on water quality not found in grab sample efforts. Recognizing the advantage of real-time monitoring, certain problems may be encountered when attempting to evaluate physical data in light of the anticipated biological quality of the water resource. Of particular concern is how to interpret complex multivariate physical data in a manner that will realistically reflect biological response and ecological resiliency. A problem typically confronted is how to account for conditions not measured, that is, antagonistic, additive, or possibly synergistic effects not deduced from physical monitoring alone. Since increasing evidence supports the observation that biological systems serve as integrators of the environmental factors upon which they depend and to conditions to which they are subjected, and realizing that their responses to interacting factors in complex ecological systems are difficult, if not impossible, to predict from physical/chemical measures alone, real-time biosensing of compensating behavioral and physiological responses has wide appeal [2,3].

In meeting the objective of developing an automated, multipurpose biosensing unit flexible enough for use in remote stream acidification monitoring, emphasis was placed on system design, hardware development and selection, data management, and field testing. System design and hardware selection was based upon availability and performance criteria of components needed in meeting the requirements of a 5-year stream acid precipitation mitigation program (APMP) in the Southern Appalachians [4].

Nature of the Problem

Background

Leading to the development of an automated fish breathing response detector having sufficient design flexibility for interacting with remote water quality data collection platforms (DCP) and communicating data via earth satellite, various design options needed to be addressed; bench testing had to be done, and those options selected for further consideration had to be subjected to the rigors of field trials. Coupled to technical requirements, specific biological concerns had to be evaluated. Input from these efforts would then be used to develop a strategy for directing remote biosensing operations involving instream fish-holding chamber design, hookups, and the detection and measurement of fish breathing responses. Once completed, biosensing devices had to be interfaced to DCPs and bench tested, and optimal configurations had to be verified under field applications before further progress could be expected.

This sequence of events would be used to meet the objective of designing similar systems for a long-term mitigative liming demonstration project planned for a high-elevation stream sensitive to acid deposition processes.

Remote Biosensing Strategy

Specially designed biosensing chambers constructed from 10 by 30-cm plastic pipe for holding individual test fish were horizontally attached to a rack or retaining device and positioned in-stream where fish were exposed to constant flow-through conditions. The relationship of fish to chamber size was maintained at approximately 1:5, with adequate space for the test subject to accommodate to existing flows. Fish were confined within test chambers by nylon screen caps attached to alternate ends of the device. Two stainless steel probes extending vertically the diameter of the chamber were connected via coaxial cables to individual amplifiers, allowing detection of opercular voltage differentials. Low-noise differential amplifiers, which produced an eventual gain exceeding 5×10^5 times the initial voltage, were designed specifically for soft-water streams with low specific conductance. Amplifier gain and amplitude trigger levels were adjustable to compensate for site-specific variations in water quality characteristics. Thus, fish breathing activity was amplified to give a representative analogue voltage of adequate amplitude or converted to a digital pulse prior to processing by the interface microprocessor-controller, in these particular applications by cable to streamside DCPs [2,3].

Interfacing Remote Data Collection Platforms

Signals generated by individually monitored fish were received by the DCP in analogue voltage or digital form, depending on the particular interface used [2]. The data were held for any number of preset monitoring intervals by interface counters or directly by the programmed DCP. Data presented to the DCP were transmitted to the NOAA Geostationary Operational Environmental Satellite (GOES) on six occasions each day or held by the DCP to be transmitted during a specified satellite window. Broadcast data received by satellite were retransmitted to a data coordinating processing center (Fig. 1). Depending on site-specific needs, simultaneous transmission of water quality parameters included: temperature, dissolved oxygen (DO), hydrogen ion concentration (pH), specific conductance, oxidation-reduction (redox) potential, and stage height. Physical sensors were positioned in-stream alongside fish-holding chambers.

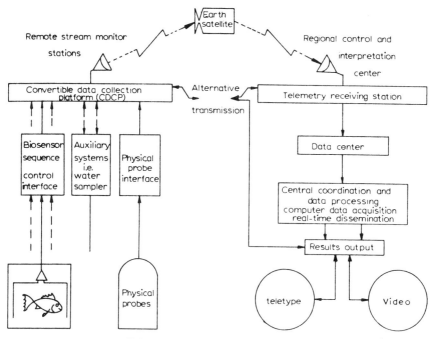

FIG. 1—*Remote water quality monitoring.*

Prototype Field Trials

Two remote biosensing systems were developed, each designed to meet specific requirements and objectives of the drainage selected. Prototype systems were tested at a warm-water site on the Cumberland River near Carthage, Tennessee, and on cold-water streams in the Great Smoky Mountains National Park.

In the Cades Cove basin of the Great Smoky Mounains National Park, biosensing devices were established at two separate water quality monitoring stations on Abrams Creek, each equipped with a DCP. We chose to take this opportunity to test two types of interface units that had been designed for different modes of data input from the biosensor. One prototype unit was designed for interfacing to the serial port of the upstream DCP and provided breathing rate data in analogue voltage ranging from 0 to 5 V. At the downstream site, a different type interface was tested in the parallel buffered input mode. This interface provided the advantage of continuously storing breathing events, where they were accumulated in counters for preset monitoring intervals and submitted as digital data to the DCP. At both stream stations, rainbow trout (*Salmo gairdneri*) were positioned in individual holding chambers and the remote device interfaced to streamside water quality platforms. Sensors from accompanying physical monitors were positioned in-stream alongside the fish chambers, and, during each DCP update, simultaneous readings for selected physical parameters were taken, that is, temperature, DO, pH, and conductivity.

The Cumberland River DCP site was located along the south shore under the Cordell Hull Bridge. In establishing remote biosensing devices at this station, fish-holding chambers containing sunfish (*Lepomis macrochirus* or *L. cyanellus*) were attached to a rack anchored to the base of the concrete bridge support nearest the south shore. Coaxial cables attached to probes fixed to three fish chambers were connected to amplifiers levied in a waterproof container midway up the 30-m-high support structure, which in turn were extended to a substation located on top of the bridge support. Here, using a battery-power source, digital signals received from the amplifiers were boosted across 154 m of cable to the DCP and accompanying monitoring equipment on shore. In this configuration, fish breathing rate data would be received at each DCP update of 73 s. Simultaneous measurements of temperature and stage height were taken for this particular application. An update was programmed for each 15-min interval with the data stored from each update in the DCP. The information in its entirety was transmitted to the satellite every 4 h [4].

Acid Deposition Mitigation Monitoring

In 1980 Congress established the National Acid Precipitation Assessment Program (NAPAP) with a 10-year renewable mandate to increase our understanding of the causes and effects of acid deposition. Of the ten working groups specified by a designated interagency task force, the aquatic effects group identified five objectives, including the "development of mitigative strategies for restoration of acidified lakes and streams." As participants in NAPAP efforts with mandated responsibilities for protecting and enhancing the Nation's fish and wildlife resources, the Eastern Energy and Land Use Team, under the U.S. Fish and Wildlife Service, has supported several 5-year cooperative field research programs through state agencies to develop and evaluate acid deposition mitigation strategies for acidified watersheds [5].

In Tennessee, the Department of Health and Environment has accepted primary responsibilities for identifying and evaluating ecological responses to mitigative lime treatments in Laurel Branch creek, a high elevation stream in the Southern Appalachians believed to be influenced by acidification processes [6,7]. Participating with the state agency, U.S. Tennessee Valley Authority (TVA) and Tennessee Technological University, with cooperation from the U.S. Forest Service and Tennessee Wildlife Resource Agency, have assisted in technical and logistical aspects of the program.

Coupled with evaluating benthic macroinvertebrate community structures, trout population dynamics, and an exhaustive series of water chemistry characteristics, two remote water quality DCPs have been positioned along the 4.3 km length of Laurel Branch, a tributary to North River in the Tellico River drainage of Tennessee.

Remote Computer-Assisted DCP

A general-purpose, self-contained computer which functions as a central controller and processor in conjunction with input/output and various communication modules was selected from a commercial vendor (Model 3400 remote computer, Synergetics International) for use at the remote stations. Features include expandable PROM/RAM memory, a programmable calendar/alarm clock, an S-34 controller port, RS-232 operator port, and being supported by the S-FORTH operating system. Based on the FIG-FORTH model, S-FORTH has been extended to include real-time, multitasking features and is a stack-oriented, thread-interpretive language.

Other advantages of the system's functional modularity are options in communication mode and end-device interface modules. For input information, signal conditioning is incorporated, allowing sensors, instruments, and actuators to be wired directly to the specific module without the expensive and time-consuming interface development required of earlier platforms. The input/output channels are lightning and surge protected, a feature not found in platforms used in earlier field trials. Peripheral expansion or new applications are easily implemented through the S-34 bus, which uses a true parallel architecture and is also configured with the complementary-symmetrical metal oxide semiconductor (CMOS) family of integrated circuits (IC). This general-purpose intermodule communications standard or "data highway" has the expansion flexibility of supporting up to 255 modules linked in analog and digital domains using a single 30.5-m (100-ft) flat ribbon cable. A major improvement is the inclusion of the positive transaction feedback support, provided by the transfer strobe line, which adds increased reliability over the daisy chains previously encountered. In addition to the satellite communication links we are employing, landline, radio frequency, and microwave transceivers may be utilized singly or in combination for redundancy. The only backup system presently being considered in our remote application is magnetic tape recorders. Physical sensors and instrumentation for stream temperature, specific conductance, pH, and flow are modules developed for remote water quality monitoring by the U.S. Geological Survey. These devices are directly compatible with the DCP input mode.

Fish Breathing Rate Detectors

Configured on printed circuit boards, breath detector amplifiers feature a stable differential input amplifier, filter and gain operational amplifiers, and a signal level detector. Each trout breath is sensed by detection probes, filtered to reject noise and interference, and amplified more than 500 000 times. The modulated signal then is converted to a digital signal for transmission to the accumulators in the interface. Two capacitors couple the input probes to the instrumentation amplifier, preventing amplification of director current potentials between probes. Initial amplifier gain is set to 1000 by precision resistors, and the output-offset voltage may be adjusted by "shorting" the amplifier inputs and adjusting an onboard potentiometer. Utilizing a quad-operational IC amplifier, the signal is passed from the input to one of four amplifiers serving as a double-pole, low-pass filter. A second amplifier of the IC is coupled, providing constant gain. The gain of the third amplifier in this IC is adjusted by a potentiometer, while the fourth amplifier in the IC is not used in this configuration. Voltage level detection with hysteresis also is provided in the circuit. When the analog voltage signal rises above +5 Vdc, the output switches to a high level, causing the accumulator in the interface to be incremented. The output will switch from a high to a low state when the analog signal drops below 2 Vdc. Line drivers are mounted on a separate board in the amplifier case to ensure reliable transmission.

In-Stream Fish Chambers and Stream-Side Facilities

Holding chambers designed for rainbow trout employ the same plastic-tube design tested in earlier field trials except that the paired probes used in detecting breath responses have been positioned along the horizontal length of the chamber [3,8]. Utilizing various computer-assisted biomonitoring systems, this chamber probe configuration was found to be superior to other designs, not only for fish, but for small chambers designed for the burrowing mayfly, *Hexagenia* spp.

Eight fish are positioned in situ at each of the two sites: one group upstream above the liming device serving as a reference and another downstream at the mitigated site just above the stream's confluence to North River. At streamside, metal fabricated buildings house electrical equipment and instrumentation. Separate 12 Vdc heavy duty marine batteries provide current for the uninterrupted power supply needed for the DCP and transceiver, the physical sensor instrumentation, and the fish breathing rate detector. The motor-driven liming device also has a separate 12 Vdc supply. Though photoelectric recharging is used, batteries are checked monthly and replacements made as needed.

Central Data Receiving and Processing

As remote stations become operational, digitally encoded information for in situ fish breathing rates and physical water quality measurement will be transmitted at ultra-high frequencies via satellite to the central data collection center located in TVA's data management facilities at Knoxville, Tennessee. Operating through the GOES, real-time stream data will be available to any user via personal computers at 3-h updates. Data files and analysis will be managed by TVA's data services branch, the state agency, and the university.

Summary

Fish physiological and behavioral responses have been measured in the laboratory as a means of observing cause/effect relationships resulting from acidification processes. Providing real-time, dose-dependent stress responses of rainbow trout subjected to stream acidification events could provide an important complement to acid deposition mitigation programs. In meeting this objective, we developed and tested different components and configurations of an automated fish gill ventilatory (breathing) rate detector for application to remote computer-assisted, data collection platforms equipped with satellite communications. By coupling these technologies to a 5-year lime mitigative program, biological and physical water quality data are available for management needs on a real-time basis.

Acknowledgments

Support for the developmental phase of this project was provided by funds made available to the Tennessee Water Resources Research Center by the following participants: Environmental Science Division; Oak Ridge National Laboratory; Office of Natural Resources, U.S. Tennessee Valley Authority; and Office of Water Research and Technology, U.S. Department of the Interior (USDI). Additional funding was provided by the U.S. Corps of Engineers, Nashville District, and the Office of Research, Aquatic Ecology Fund, Environmental Biology Research Program, Tennessee Technological University. Ongoing support is made available through USDI, Fish and Wildlife Service; Eastern Energy Land Use Team, via Tennessee Department of Health and Environment; and the Center for Management, Utilization, and Protection of Water Resources, Tennessee Technological University.

References

[1] *Biological Monitoring in Water Pollution,* Cairns, J., Jr., Ed., Pergamon Press Ltd., Oxford, England, 1982.
[2] Morgan, E. L., Eagleson, K. W., Herrmann, R., and McCollough, N. D. *Journal of Hydrology,* Vol. 51, No. 4, 1981, pp. 339–345.
[3] Morgan, E. L. and Young, R. C., *Verh. International Verein. Limnology,* Vol. 22, 1984, pp. 1432–1435.
[4] Morgan, E. L. and Eagleson, K. W., "Automated Biomonitoring Applications in Remote Water Quality Surveillance and Time Rate Toxicological Assay," Technical Report No. 86, Tennessee Water Resources Research Center, Knoxville, Tennessee, 1982.
[5] Saunders, W. P., Britt, D. L., Kinsman, J. D., DePinto, J., Rogers, P., Effler, S., Sverdrup, H., and Warfringe, P., "APMP Guidance Manual—Vol. I: APMP Research Requirements," U.S. Fish and Wildlife Service, Eastern Energy and Land Use Team, Biol. Rep. 80(40.24), 1985, p. 207.
[6] Morgan, E. L., Porak, W. F., and Arway, J. A. "Controlling Acid-Toxic Metal Leachates from Southern Appalachian Construction Slope: Mitigating Stream Damage," in *Wetlands, Floodplains, Erosion, and Storm Water Pumping,* Transportation Research Board Record 948, No. 86, 1983, pp. 10–16.
[7] Olem, H., "Acidification Trends in Surface Waters of the Southern Appalachians, Final Report," U.S. Tennessee Valley Authority, TVA/ONRED/AWR—85/13, 1985, p. 54.
[8] Morgan, E. L., Young, R. C., and Crane, C., "Automated Multi-Species Biomonitoring Employing Fish and Aquatic Invertebrates of Various Trophic Levels," in *Freshwater Biological Monitoring,* D. Pascoe and B. W. Edwards, Eds., Pergamon Press, Oxford, England, 1984, pp. 75–80.

Richard C. Young[1]

Consideration of Total Ion Composition in Designing Toxicity Tests Using Aluminum Salts and Mineral Acids

REFERENCE: Young, R. C., "**Consideration of Total Ion Composition in Designing Toxicity Tests Using Aluminum Salts and Mineral Acids,**" *Impact of Acid Rain and Deposition on Aquatic Biological Systems, ASTM STP 928,* B. G. Isom, S. D. Dennis, and J. M. Bates, Eds., American Society for Testing and Materials, Philadelphia, 1986, pp. 92–97.

ABSTRACT: Review of the current literature reveals that many researchers involved in the study of the effects of aluminum and pH on aquatic life are using aluminum salts, that is, sulfate or chloride forms, to obtain the desired aluminum concentration for their experiments. This technique raises some concern for soft water experiments due to the unknown effect of excessive concentrations (in relation to background) of the associated anions $SO_4^=$ or Cl^-. For example, if aluminum chloride ($AlCl_3$) is employed to obtain a concentration of 0.5 mg/L aluminum, the solution also will be increased by 2.0 mg/L Cl^-. Likewise, if aluminum sulfate [$Al_2(SO_4)_3$] is used to achieve the 0.5 mg/L aluminum, $SO_4^=$ will be increased 2.6 mg/L. For inland streams (that is, not affected by ocean salts), these values appear to be out of proportion. In some instances the use of aluminum salts can easily double or increase by the power of 10 the concentrations of associated anion. $SO_4^=$ or Cl^- concentrations are further increased by the use of sulfuric and hydrochloric acid to adjust pH. Since aluminum is an effective buffer below pH 5.0, solutions containing high concentrations of aluminum also will require larger volumes of acid to achieve pH values below this level, thus compounding the problem.

The question of acid form, that is, sulfuric, hydrochloric, nitric, etc., in relation to toxicity leads to the consideration that these "strong acids" are only one component contributing to the total acidity of a system. In coniferous forests with deep acid soils, storm water runoff could contribute significantly to the acidity of streams due to fulvic acids. Their chelating ability also could enhance the mobilization of aluminum. How these weak organic acids affect or regulate pH or pH/aluminum toxicity in fish is also unknown.

The paper discusses the importance of total solute concentration and the possible role specific anions play in mediating osmoregulation in acid- and acid/aluminum-stressed fish.

KEY WORDS: acid precipitation, fish bioassay, aluminum, pH, techniques

The relationship between the solution pressures of metal salts and their toxicity to fish was initially investigated by Jones [1], who demonstrated that the toxicity of a metal was, in general, inversely related to its solubility. Jones noted that solutions of metal salts of iron and aluminum were particularly distinctive because

[1] Principal, The ADVENT Group, Inc., P.O. Box 1147 Brentwood, TN 37027.

their solutions were acidic yet more toxic than solutions of equivalent pH prepared from strong acids. Aluminum was shown to be most toxic at molar concentrations of aluminum sulfate which yielded a pH near 5.0.

In reviewing the current literature, it is noted that many researchers involved in the study of the effects of aluminum and pH on aquatic life are using aluminum salts, that is, sulfate or chloride forms, to obtain the desired aluminum concentrations for their experiments. For most inland waters, the addition of a few milligrams per liter of Cl^- or $SO_4^=$ in the course of conducting metal toxicity tests would be of little concern. However, employing this technique in waters characteristically dilute in ionic strength such as those associated with acid precipitation aquatic effect studies raises some concern due to the unknown effect of excessive concentrations (in relation to background) of the associated anions $SO_4^=$ or Cl^-. For example, if aluminum chloride ($AlCl_3$) is employed to obtain a concentration of 0.5 mg/L aluminum, the solution also will be increased by 2.0 mg/L Cl^-. Likewise, if aluminum sulfate $[Al_2(SO_4)_3]$ is used to achieve the 0.5 mg/L aluminum, $SO_4^=$ will be increased 2.6 mg/L. For Appalachian streams (that is, not affected by ocean salts), these values appear to be out of proportion. For instance, baseline Raven Fork Creek (TVA acid precipitation study site near Cherokee, North Carolina) Cl^- concentrations range from 0.5 to 1.0 mg/L and $SO_4^=$ concentrations 0.4 to 1.6 mg/L [2]. Maximum concentrations during storm water flow are 1.0 mg/L Cl^- and 3.0 mg/L $SO_4^=$ [3]. Values for annual average volume weighted rainfall near the site (Blount County, Tennessee) in 1980 were 0.24 mg/L Cl^- and 2.5 mg/L $SO_4^=$ [4]. In contrast, Schofield and Trojnar [5] report that high elevation Adirondack lakes typically have Cl^- concentrations between 0.2 and 0.7 mg/L and $SO_4^=$ concentrations between 3 and 11 mg/L. Therefore, substantial differences exist between high elevation lakes of the northeast and mountain streams of the Southern Appalachians. The chemistry is further complicated by the fact that in many of these systems the acid fraction may be dominated by weak organic acids. Physiological responses (toxicity) of fish to organic acids alone is poorly understood.

Therefore, in some instances the use of aluminum salts can easily double or increase by a power of ten the concentrations of associated anions. $SO_4^=$ or Cl^- concentrations are further increased by the use of sulfuric and hydrochloric acid to adjust pH. Since aluminum is an effective buffer below pH 5.0, solutions containing high concentrations of aluminum also will require larger volumes of acid to achieve pH values below this level.

Discussion

Most researchers are in agreement that pH and pH/aluminum stress in fish result in depletion of body salt content, hyperventilation, and decreased blood oxygen-carrying capacity. Muniz et al. [6], Leivestad et al. [7], Muniz and Leivestad [8], and Nelson [9] provide pertinent discussions describing the physiological effects of pH and pH/aluminum stress on fish. Also, Hoar and Randall

[10] and Krogh [11] give detailed mechanisms for osmoregulation and blood electrolyte balances for fish in low ionic strength waters. However, these papers do not address how anions, such as Cl^- or $SO_4^=$, affect the osmoregulatory process under stress conditions. Since electrolyte loss is a key response, it is possible that this increase in anion concentration may act to reduce the loss of body electrolytes because it would decrease the magnitude of the osmotic pressure differential between the external medium and the internal body fluid. Such action would effectively be antagonistic. In this relation, $AlCl_3$ would be twice as effective in increasing solution osmotic pressure as sodium chloride (NaCl) (on a mole per mole basis) since $AlCl_3$ produces four ions and NaCl produces only two (for example, $p = iCRT$, where p = osmotic pressure, i = number of ions formed, C = molar concentration, R = gas constant, and T = absolute temperature).

Few studies are reported in the literature on the differential effect of SO_4^-. However, Garcia-Romeu and Maetz [12] have demonstrated that $SO_4^=$ does not exchange across or affect gill membrane permeability as does Cl^-. Graham and Wood [13] concluded that $SO_4^=$ at pH $\geq$ 4.0 acted as a potentiator of other stresses to rainbow trout in soft waters. They also concluded that sulfuric acid (H_2SO_4) was generally less toxic to rainbow trout than hydrogen chloride (HCl) and that exhaustive exercise markedly increased the toxicity of H_2SO_4, but not HCl.

In bog systems, several species of fish exist and propagate in waters which have pH values of 4.0 to 4.5 [14]. Baker [15] suggests that fish existing at low pH in bog drainages are genetically adapted and also protected from aluminum toxicity by organic acid chelation. Hermond [16] studied the systems and showed that pH (3.8) was maintained by organic acids and that the sulfate reduction and nitrate uptake were effective buffers of acidic atmospheric loadings. Glover and Webb [17] found that 60% of the hydrogen concentration in the Tovdal River was contributed by weak acids originating from the soil. Henriksen and Seip [18] concluded that aluminum, silica, and fulvic acid were the major components contributing to weak acids in lakes in Southern Norway and Southwestern Scotland, and that variance in weak acid concentrations was explained by their organic carbon and aluminum content (inorganic aluminum contributed as much as 50% of the weak acidity and each milligram of organic carbon was equal to 5.5 μeq weak acid). Schnitzer [19] shows that the major organic acids in water are derived from humic substances in the soil and are composed of humin, fulvic, and humic acids, fulvic acid being commonly the most dominant.

Fulvic acids also are known to be significant modifiers of metal ion chemistry. Saar and Weber [20] describe three important concerns relating to fulvic acids in organic systems: (1) metal ions are known to be less toxic to aquatic organisms when complexed with ligands, such as fulvic acid, and, consequently, in toxicity studies, complexed and hydrated metal species must be considered separately; (2) fulvic acid can alter the geochemical mobility of metal ions in the aquatic environment and act essentially as a metal ion buffer; and (3) fulvic acids increase observation that fulvic acid metal ion complexes are most dominant at intermediate pH levels, that is, H^+ competes with metals for anionic bonding sites while OH^- competes with fulvic acid for cationic metals. Therefore, as the pH is lowered the partition coefficient changes and metals may be released.

In an effort to estimate the effectiveness of humic acids in reducing the toxicity of aluminum to early life stages of fish, an investigation by Driscoll et al. [21] employed citrate as a surrogate for humate and concluded that aluminum complexed with citrate was not toxic. However, this observation must be interpreted cautiously because natural organic acids are very complex macromolecules with the capacity of chelating aluminum by several bonding configurations which are much weaker than that formed by citrate. Possibly, under natural conditions, aluminum chelated by organic acids is released by transient pH changes and becomes toxic to fish. The hypotheses that organic acids are less toxic than mineral acids and that organically bound aluminum is not toxic to fish deserve further examination.

Another important factor which has received little attention is the influence of temperature on the toxicity of pH and inorganic aluminum. We expect that systems undergoing acidification which are also subject to large seasonal or transient temperature changes may be especially harmful to fish populations. The effect of temperature on the reactions of aluminum in dilute waters is not well documented (Driscoll, personal communication) and, since the toxicity of aluminum to fish seems to be dependent on ionic form [19], temperature could be significant. The effect of temperature on toxicity and metabolism per se is fundamental, since elevation in temperature shifts the blood oxygen dissociation curve to the extent that the half-saturation oxygen pressure (P_{50}) is elevated, making it more difficult for the blood to unload carbon dioxide (CO_2). Hemoglobins of vertebrates, including fish, are highly variable among species and, in this response, it may be predictable that fish species highly efficient in gas exchange are more tolerant of acid stress. For example, fish such as trout, which live in well-aerated waters, are more responsive to decreases in partial pressure of oxygen (PO_2) or increases in partial pressure of CO_2 (PCO_2) than many of the more resilient warm water fishes. With respect to pH effects alone, organic acids may not be as potent at the same hydrogen ion concentration as strong mineral acids. This hypothesis, namely the effect of these large molecular weight polyelectrolytes on gill membrane permeability and hydrogen ion toxicity, is an area of needed research.

This discussion supports the concern that total ionic composition plays an important role in mediating osmoregulation in acid- and acid/aluminum-stressed fish. Which ions are more important and at what concentrations they become effective is not well understood. The hypothesis deserves further thought to ensure that laboratory pH and pH aluminum toxicity studies are providing meaningful data. Brown [22] states that Wright and Snekvik [23] found no correlation between lake aluminum concentration and fishery status. This is contrary to Brown's laboratory results, which showed that the aluminum concentrations observed by Snekvik were in the range that should have adversely affected the fish populations.

The difference in the observed and predicted results were explained in part by the fact that observed lake data for aluminum were total aluminum, and the laboratory data were developed from inorganic aluminum and, therefore, much of the total aluminum from the field data was presumably complexed and nontoxic.

Such results further emphasize the need for laboratory studies which mimic natural ionic compositions as closely as possible. Experiments should, therefore, be sensitive to adding aluminum in a form which is chemically representative of the specific aquatic environment being studied.

To avoid this potential problem, the following alternative laboratory protocol for preparing solutions of aluminum for toxicity studies is suggested. Aluminum hydroxide precipitate is prepared by dissolving the appropriate amount of $AlCl_3$ (hexahydrate) in distilled water. The $AlCl_3$ solution is then titrated with sodium hydroxide (NaOH) to pH $\cong$ 7. Aluminum hydroxide is precipitated as a fine white flock. In general the reaction is

$$AlCl_3 + NaOH \xrightarrow{pH \cong 7} Al(OH)_x + NaCl$$

The NaCl is removed from the precipitated solution by dialysis. The freshly precipitated salt-free aluminum hyroxide solution then is used to spike the test solution at the desired concentration. This procedure allows for the preparation of aluminum hydroxide that mimics that of the natural dissolution of gibbsite, that is,

$$Al_2O_3(H_2O)_3 + 2 H_2O \leftrightarrows 2 Al(OH)_4^- + 2 H^+$$

but in larger concentrations and less time. Freshly formed aluminum hydroxide [$Al(OH)_x$] also has the property of reacting quickly (when compared with the dissolution reactions of the mineral aluminum) to change species/form with respect to solution pH.

Future research needs include investigation of the role of organic acids in mediating aluminum chemistry and toxicity to aquatic biota. This may require more effort in developing laboratory protocols (such as described for aluminum) for mimicking the chemical composition of acid-sensitive aquatic environments than current research is providing.

Acknowledgment

This paper was prepared while the author was employed by the Tennessee Valley Authority, Division of Air and Water Resources. The Agency support and technical review by the staff at the Aquatic Environmental Research Laboratory are gratefully acknowledged.

References

[1] Jones, J. R. E., "The Relation Between the Electrolytic Solutions Pressures of the Metals, Their Toxicity to the Stickleback (*Gasterosteus aculeatus L.*)," *Journal of Experimental Biology*, Vol. 16, 1939, pp. 425–437.

[2] "Mineral Quality of Surface Waters in the Tennessee River Basin," Tennessee Valley Authority, Hydraulic Data Branch, Knoxville, TN, 1972.
[3] "Raven Fork Acid Precipitation Study," Tennessee Valley Authority, Water Quality Branch, Chattanooga, TN, 1982a, unpublished data.
[4] "Regional Air Quality Monitoring Network," Tennessee Valley Authority, Air Quality Programs, Muscle Shoals, AL, 1982b, unpubished data.
[5] Schofield, C. L. and Trojnar, J. R. in *Polluted Rain*, T. Y. Toribara, N. W. Miller, and P. E. Morrow, Eds., Plenum Press, NY, 1980, pp. 341–366.
[6] Muniz, I. P., Leivestad, H., and Rosseland, B. O., "Stress Measurements on Fish in Acidic Rivers," *Nordforsk, Publ.*, Vol. 2, 1978, pp. 233–247.
[7] Leivestad, H., Hendry, G., Muniz, I. P., and Snekvik, E., "Effects of Acid Precipitation on Freshwater Organisms," in *Impact of Acid Precipitation on Forest and Freshwater Ecosystems in Norway*, F. H. Braekke, Ed., SNSF project, FR6/76, 1976, pp. 87–111.
[8] Muniz, I. P. and Leivestad, H., "Toxic Effects of Aluminum on Brown Trout, *Salmo trutta L*," in *Ecological Impact of Acid Precipitation*, D. Drablos and A. Tollan, Eds., SNS project, 1980, pp. 320–321.
[9] Nelson, J. A., "Physiological Observations on Developing Rainbow Trout, *Salmo gairdneri* (Richardson), Exposed to Low pH and Varied Calcium Ion Concentration," *Journal of Fish Biology*, Vol. 20, 1982, pp. 359–372.
[10] *Fish Physiology*, Vol. I: *Excretion, Ionic Regulation, and Metabolism*, W. S. Hoar and D. J. Randall, Eds., Academic Press, New York, 1969.
[11] Krogh, A., *Osmotic Regulation in Aquatic Animals*, Dover Publications, Inc., New York, 1965.
[12] Garcia-Romeu, F. and Maetz, J., "The Mechanism of Sodium and Chloride Uptake by the Gills of a Freshwater Fish, *Carassius auratus:* I. Evidence for an Independent Uptake of Sodium and Chloride Ions," *Journal of General Physiology*, Vol. 47, 1964, pp. 1195–1207.
[13] Graham, M. S. and Wood, C. M., "Toxicity of Environmental Acid to Rainbow Trout: Interactions of Water Hardness, Acid Type, and Exercise," *Canadian Journal of Zoology*, Vol. 59, 1981, pp. 1518–1526.
[14] Rahel, F. J. and Magnuson, J. J., "Fish in Naturally Acidic Lakes," in *Proceedings*, Conference on the Ecological Impact of Acid Precipitation, SNSF project report, Oslo, Norway, 1980.
[15] Baker, J. P., "Effects on Fish of Metals Associated with Acidification," in *Acid Rain/Fisheries*, T. A. Haines and R. E. Johnson, Eds., American Fisheries Society, Bethesda, MD, 1982, pp. 165–176.
[16] Hermond, H. F., "Biogeochemistry of Thoreau's Bog," *Ecological Monographs*, Vol. 50, No. 4, 1980, pp. 507–526.
[17] Glover, G. M. and Webb, A. H., "Weak and Strong Acids in the Surface Waters of the Tovdal Region in Southern Norway," *Water Research*, No. 13, 1979, pp. 781–784.
[18] Henriksen, A. and Seip, H. M., "Strong and Weak Acids in Surface Waters of Southern Norway and Southwestern Scotland," *Water Research*, Vol. 14, 1980, pp. 809–813.
[19] Schnitzer, M., "The Chemistry of Humic Substances," *Environmental Biochemistry*, J. O. Nriagu, Ed., Ann Arbor Sciences Publishers, Michigan, 1976.
[20] Saar, R. A. and Weber, J. H., "Fulvic Acid: Modifier of Metalion Chemistry," *Environmental Science Technology*, Vol. 16, No. 9, 1982, pp. 510a–517a.
[21] Driscoll, C. D., Baker, J. P., Bisogni, J. J., and Schofield, C. L., *Nature*, Vol. 161, 1980, p. 284.
[22] Brown, D. J. A., "Effect of Calcium and Aluminum Concentrations on the Survival of Brown Trout (*Salmo trutta*) at Low pH," *Bulletin of Environmental Contamination and Toxicology*, May, 1983.
[23] Wright, R. F. and Sekvik, E., *Verh. Internat. Verein. Limnol.*, Vol. 20, No. 765, 1978.

Peter F. Boyle,[1] James W. Ross,[1] John C. Synnott,[1] and Cheryl L. James[1]

A Simple Method to Measure pH Accurately in Acid Rain Samples

REFERENCE: Boyle, P. F., Ross, J. W., Synnott, J. C., and James, C. L., "**A Simple Method to Measure pH Accurately in Acid Rain Samples,**" *Impact of Acid Rain and Deposition on Aquatic Biological Systems, ASTM STP 928,* B. G. Isom, S. D. Dennis, and J. M. Bates, Eds., American Society for Testing and Materials, Philadelphia, 1986, pp. 98–106.

ABSTRACT: Conventional pH electrodes, which are designed to function in high-conductivity solutions, show slow and erratic response in pure water samples such as acid precipitation. Increasing conductivity without shifting pH solves the measurement problem. This method correctly measures pH in low conductivity solutions, using a Ross combination pH electrode, after the addition of a small amount of potassium chloride (KCl) solution to the sample. The Ross electrode, with its unique internal redox system, is chosen since it eliminates problems associated with temperature and because the precision and accuracy of data obtained in preliminary testing with it were comparable to those obtained with the standard hydrogen electrode. KCl addition to samples does not alter the pH significantly.

KEY WORDS: pH, acid precipitation, low conductivity, Ross electrode, standard hydrogen electrode

As concern for the effects of acid precipitation on aquatic biological systems has grown, a considerable body of pH and alkalinity data has been generated. Such information appears regularly in the media, serves as the basis for new legislation, and continues to be the focus of litigation. The measurements are usually made electrometrically, using pH and reference electrodes, with calibration against standard buffers. Because these devices have been in use for so long, there is a tendency to extend their application without adequate recognition of their operational limitations. Moreover, acid precipitation samples, in their relative purity, are subject to easy contamination from atmospheric carbon dioxide, from sample containers, and from sample-handling techniques. Temperature and solution carryover also would impact significantly on such fragile samples.

Unfortunately, conventional pH measurement techniques have been widely applied to these samples. This may well mean that the accuracy and reproducibility

[1] Research scientist; vice-president, Research; director, Applications Research; and research chemist, respectively, Orion Research, Inc., Cambridge, MA 02139.

of a substantial amount of pH data cannot be relied upon, and, worse yet, that unreliable data continue to be generated. In this work, we have examined a number of variables that affect measurement of pH in low conductivity, high purity waters and have arrived at a method that addresses them (see Appendix). We are sufficiently confident of the method's utility that we are introducing it into the ASTM standards process for collaborative testing.

The Problem

The major difficulty in the measurement of pH in acid precipitation can be attributed, ironically, to the purity and therefore, very low conductivity of the samples. Conventional pH electrodes are designed to function in highly conductive solutions and will respond more slowly and erratically as conductivity decreases. The effect also is felt at the reference electrode, where changes in liquid junction potential, caused by variations in ionic strength, can markedly alter measured values. Solution carryover from calibration buffers, another aspect of the same problem, also can lead to erratic results.

Attempts have been made to remedy the situation through the construction of electrodes with low-resistance sensing elements in the belief that higher conductivity glass would promote faster response and improved stability. Indeed, such "pure water pH electrodes" would be expected to show reduced noise levels. However, when these electrodes were examined for use in measuring pH in boiler feed waters, Midgley and Torrance concluded that "low-resistance electrodes had no significant advantages over general purpose electrodes for the application to power station waters, although they were capable of slightly higher precision." Approaches which try to minimize junction potential excursions can result in nonreproducible sample contamination from the reference electrode filling solution. All of these approaches are adaptions of the method to accommodate sample conditions, and their failure to solve the problem led us to consider adjusting the samples to meet measurement requirements. We hoped, by addition of a neutral salt to the sample, to increase conductivity without shifting pH. Conventional measurement techniques then could be applied.

TABLE 1—*Conventional pH measurement of low conductivity samples.*

Electrode Designation	Sample pH
C-1	5.80
C-2	7.05
C-3	6.30
C-4	5.75
C-5	5.41
C-6	5.90
C-7	5.21
Standard hydrogen	5.78

TABLE 2—*Effect of stirring and glass resistance on pH measurement.*

HCl, M	Actual pH[a]	High-Resistance Glass		Low-Resistance Glass	
		Stirred	Unstirred	Stirred	Unstirred
10^{-6}	6.73	4.63	4.55	5.09	4.68
10^{-5}	5.31	4.75	4.40	4.86	4.40
10^{-4}	3.99	4.06	3.80	4.08	3.78
10^{-3}	2.97	3.09	3.06	3.08	3.06
10^{-2}	1.98	2.12	2.11	2.12	2.12
10^{-1}	1.03	1.36	1.30	1.35	1.41

[a] As determined using standard hydrogen electrode.

In order to establish a basis for validating our method, we first tried to ascertain the magnitude of the problem. We isolated a supply of distilled, deionized water, protected with a carbon dioxide trap, sufficient for all anticipated testing. Seven different commercially available electrodes were obtained, and a standard hydrogen gas electrode was prepared to serve as referee in the testing. Each electrode pair was calibrated in standard buffers and was used, in conjunction with an Orion model 901 digital pH/mV meter, to determine the pH of the distilled deionized water. The results, appearing in Table 1, provide a rather dramatic statement of the problem. This was supported in subsequent evaluations of the effects of stirring and glass resistance. These tests utilized dilute acid samples, and all were compared with the standard hydrogen electrode results (Table 2). Neither stirring nor glass resistance had much effect on measurement. Experiments to this point suggested that pH measurement of dilute acid samples, using conventional techniques, provides reasonably consistent, albeit wholly incorrect information.

Method Development

Since we hoped to solve the problem by adding a neutral salt to increase conductivity, we tested to see if that would shift sample pH. Once again, using dilute acid samples and the standard hydrogen electrode, we measured pH with

TABLE 3—*Effect of KCl addition on pH accuracy.*

HCl, M	pH—Standard Hydrogen Electrode	
	0.0 M KCl	0.01 M KCl
10^{-6}	6.733	6.703
10^{-5}	5.317	5.345
10^{-4}	3.994	3.998
10^{-3}	2.969	2.970
10^{-2}	1.977	1.978
10^{-1}	1.027	1.030

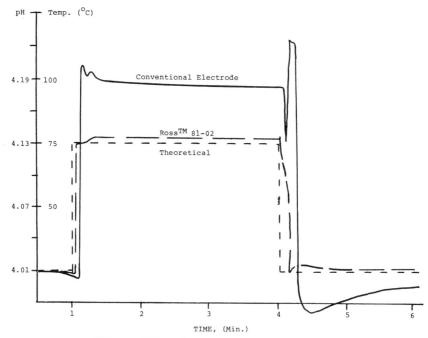

FIG. 1—*pH electrode response to temperature excursion.*

and without potassium chloride (KCl) in the background (Table 3). The effect of the salt on pH is minimal.

For development and evaluation of a method, we selected the Ross Model 81-02 combination electrode (coded C-1 in Table 1). This was based upon its good agreement with the standard hydrogen electrode and because its redox internal system makes it virtually temperature insensitive (Fig. 1). We also hoped that the free-flowing reference junction would help reduce junction potential difficulties. Under the same conditions used to establish Table 3, we verified that the addition of KCl to dilute acid samples did not affect pH as measured by this electrode (Table 4). A separate, direct comparison with the standard hydrogen electrode using similar solutions (Fig. 2) encouraged proceeding with the method.

Reference junction potential plays a major role in the measurement of pH in low conductivity samples. Differences in ionic strength between buffers and samples can cause large variations in the junction potential and, therefore, in pH values. Obviously then, contamination from buffer carryover is a real and substantial problem. Although we expected the free-flowing reference junction of the 81-02 to help alleviate the difficulty, we felt that dilution of the calibration buffers to ionic strength levels closer to sample conditions also would improve measurement precision and accuracy. To keep matters simple, we used off-the-shelf buffers at three dilutions, all spiked with 0.01 M KCl as background. The pH values were established with the standard hydrogen electrode, and then mea-

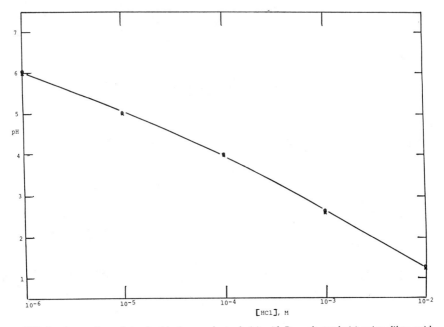

FIG. 2—*Comparison of standard hydrogen electrode (o) with Ross electrode (x) using dilute acid samples with 0.01 M KCl background.*

sured, with and without stirring, using the Ross 81-02 electrode (Table 5). These data show that stirring is not an issue, that the 81-02 electrode measures pH accurately, and that dilution results in decreased buffer capacity. They also confirm that, at these levels, buffer carryover is not a problem.

The experiments were expanded and rerun (Table 6) to include another buffer and further dilution. Clearly, the thousand-fold dilution is too large, but the overall comparability is excellent. For use in the method, we replaced the 7.00 buffer with 6.86, since the latter has the same general composition as the 7.00 and is National Bureau of Standards (NBS) traceable. For calibration, we used 1:100

TABLE 4—*Effect of KCl addition on pH accuracy.*

	pH—Ross Model 81-02 Electrode	
HCl, M	0.01 KCl	0.01 M KCl
10^{-6}	6.805	6.766
10^{-5}	5.410	5.410
10^{-4}	4.005	4.004
10^{-3}	3.001	3.001
10^{-2}	2.020	2.020
10^{-1}	1.060	1.060

TABLE 5—*Stirring effect in buffers—Ross 81-02 electrode.*

Buffer[a]	(Dilution)	Actual pH[b]	pH-Stirred	pH-Unstirred
7.00	(1:1)	6.98	6.99	6.99
7.00	(1:10)	7.22	7.23	7.23
7.00	(1:100)	7.20	7.24	7.21
4.01	(1:1)	4.01	4.00	4.00
4.01	(1:10)	4.10	4.11	4.11
4.01	(1:100)	4.36	4.38	4.37

[a] All are 0.01 M KCl.
[b] As determined using standard hydrogen electrode.

dilutions of the buffers, all made 0.01 M in KCl. The temperature dependence of these solutions was determined and appears in Table 7.

Results

Measurement reproducibility in calibration buffers is shown in Table 8. Sample measurement reproducibility was checked by running 15 samples of distilled, deionized water, with and without KCl, for each of three Ross 81-02 electrodes. The results (Table 8) show major improvement in both precision and accuracy and demonstrate, again, that conventional pH measurement (no KCl), even with good electrodes, produces nonsensical data on low-conductivity samples. For the KCl-loaded samples, equilibration time from buffer to sample was approximately 1 min with stability of ±0.002 pH units.

Experimental work was initiated in the hope of expanding the method to include other electrodes. The commercially available electrodes were checked for accuracy and reproducibility in the calibration buffers developed for the method and in distilled deionized water. The results (Table 9) were not entirely satisfactory and

TABLE 6—*pH of calibration buffers.*

Buffer[a]	(Dilution)	pH—Standard H_2 Electrode	pH—81-02 Electrode
10.01	(1:1)	10.01	10.01
10.01	(1:10)	10.25	10.25
10.01	(1:100)	10.00	10.02
10.01	(1:1000)	8.14	8.53
7.00	(1:1)	6.98	7.00
7.00	(1:10)	7.22	7.20
7.00	(1:100)	7.20	7.22
7.00	(1:1000)	7.05	7.12
4.01	(1:1)	4.01	4.01
4.01	(1:10)	4.10	4.10
4.01	(1:100)	4.36	4.40
4.01	(1:1000)	4.76	4.83

[a] All are 0.01 M KCl.

TABLE 7—*Temperature dependence of calibration buffers with 0.01 M KCl.*

Temperature, °C	pH—Diluted 6.86 Buffer	pH—Diluted 4.01 Buffer
10	7.01	4.10
20	6.98	4.10
30	6.95	4.12
40	6.94	4.15
50	6.95	4.18
60	6.98	4.22
70	7.01	4.27
80	7.04	4.32

TABLE 8—*Reproducibility—Ross 81-02 electrodes.*

Electrode	pH			
	Diluted 6.86[a]	Diluted 4.01[a]	H_2O–No KCl[b]	H_2O–0.01 M KCl[b]
1	6.972 ± 0.015	4.342 ± 0.028	4.656 ± 0.856	5.743 ± 0.161
2	6.974 ± 0.017	4.341 ± 0.027	3.444 ± 1.099	5.747 ± 0.148
3	6.970 ± 0.021	4.340 ± 0.050	5.111 ± 0.514	5.821 ± 0.210

[a] Five samples (replicates).
[b] Fifteen samples (replicates).

TABLE 9—*Reproducibility—various electrodes.*

Electrode	pH—5 Samples Each			
	Diluted 6.86	Diluted 4.01	H_2O–No KCl	H_2O–0.01 M KCl
C-1 (Ross 81-02)	6.972 ± 0.015	4.342 ± 0.028	6.350 ± 0.310	6.068 ± 0.069
C-2	6.980 ± 0.025	4.390 ± 0.064	6.716 ± 0.092	5.928 ± 0.046
C-3	6.832 ± 0.114	4.416 ± 0.034	6.630 ± 0.504	5.636 ± 0.170
C-4	6.922 ± 0.033	4.444 ± 0.032	6.582 ± 0.085	5.718 ± 0.244
C-5	6.780 ± 0.021	4.210 ± 0.035	5.300 ± 0.400	5.780 ± 0.031
C-6	6.900 ± 0.046	4.306 ± 0.034	6.222 ± 0.125	6.034 ± 0.058
C-7	6.858 ± 0.058	4.296 ± 0.122	6.584 ± 0.078	5.824 ± 0.182

TABLE 10—*Effect of pH half cell on measurement (common reference).*

Electrode	pH	
	0.0 M KCl	0.01 M KCl
C-1	5.982	6.220
C-2	5.416	5.382
C-3	5.286	5.229
C-4	5.378	5.336
C-5	5.444	5.360
C-6	5.232	5.231
C-7	4.913	5.399

TABLE 11—*Reference half cells versus standard hydrogen electrode.*

Electrode	pH	
	0.0 M KCl	0.01 M KCl
C-1	4.749	5.006
C-2	4.888	4.882
C-3	5.178	5.098
C-4	5.095	5.079
C-5	4.677	4.999
C-6	4.549	4.899
C-7	5.651	5.659

strongly suggest that choice of electrode cannot be lightly dismissed. We have tried to isolate the problem in two ways. In the first, we measured pH using all of the pH half cells against a common sleeve-junction reference electrode (Table 10). The second series used the standard hydrogen electrode against which each of the reference half cells was measured (Table 11). Neither data set provides a clear explanation of the problem, and it appears that both electrode half cells contribute. We are continuing to examine the matter.

Conclusions

1. The method developed using the Ross Model 81-02 electrode provides a fast and accurate means of measuring pH in low-conductivity samples.

2. The use of KCl to increase conductivity improves measurement reproducibility for all electrodes tested.

3. Problems with other electrodes can be traced to either or both half cells but are, as yet, not clearly understood.

APPENDIX

Method for Measurement of pH in Low-Conductivity Samples

1. *Scope*

    1.1 This method is applicable to the determination of pH in low-conductivity samples over the pH range 0 to 14.

2. *Summary of Method*

    2.1 KCl is added to calibration buffers and to samples to increase solution conductivity.
    2.2 Calibration buffers are diluted to minimize problems with reference junction potential.
    2.3 Measurement is made with the Ross Model 81-02 combination pH electrode.[2]

---

[2] Available from Orion Research, Inc., Cambridge, MA.

3. *Apparatus*

   3.1 Ross Model 81-02 combination pH electrode.
   3.2 pH meter.
   3.3 Mixer, magnetic, with tetrafluoroethylene (TFE) fluorocarbon-coated stirring bar.
   3.4 Laboratory glassware.

4. *Required Solutions*

   4.1 6.86 buffer solution.
   4.2 4.02 buffer solution.
   4.3 Potassium chloride solution, 1.0 M. Dissolve 7.46 g reagent grade KCl in distilled, deionized water and dilute to 100 mL.
   4.4 Buffer A. Transfer 1 mL 6.86 buffer solution and 1 mL potassium chloride solution into a 100-mL volumetric flask. Dilute to the mark with distilled, deionized water.
   4.5 Buffer B. Transfer 1 mL 4.01 buffer solution and 1 mL potassium chloride solution into a 100-mL volumetric flask. Dilute to the mark with distilled, deionized water.

5. *Calibration*

   5.1 Prepare the electrode and meter according to the manufacturer's instructions.
   5.2 Calibrate the electrode according to the manufacturer's instructions except that Buffers A and B are substituted for those normally used.

6. *Sample Analysis*

   6.1 Add 1 mL KCl solution to 100 mL sample. Transfer to 150-mL beaker and stir.
   6.2 Rinse the electrode with distilled, deionized water and place the tip in the sample. Continue stirring.
   6.3 Read displayed sample pH.

Bibliography

Bates, R. G., *Determination of pH, Theory and Practice*, 2nd ed., Wiley, New York, 1973.

Bobrov, V. S., and Shul'ts, M. M., *Zhurnal Prikladnoi Khimii*, Vol. 50, No. 5, May 1977, pp. 1040–1043.

Filomena, M., Camoes, G. F. C., and Covington, A. K., *Analytical Chemistry*, Vol. 46, No. 11, Sept. 1974, pp. 1547–1551.

Ives, G. J. and Janz, D. J. G., Eds., *Reference Electrodes, Theory and Practice*, Academic Press, New York, 1961.

Midgley, D., and Torrance, K., *Analyst*, Vol. 104, Jan. 1979, pp. 63–72.

Tentative Standard TSC-3, National Committee for Clinical Laboratory Standards, Jan. 1978.

Author Index

A
Allan, J. W., 54
Allard, M., 28

B
Bates, J. M., 3
Bennett, A. D., 4
Bienert, R. W., Jr., 17
Bobba, A. G., 42
Boyle, P. F., 92
Brezonik, P. L., 67
Burton, T. M., 54

C
Clarkson, C. L., 17
Crisman, T. L., 17

D
Dennis, S. D., 3

E
Eagleson, K. W., 84

G
Garren, R. A., 17

I
Isom, B. G., 3, 84

J
James, C. L., 92
Jeffries, D. S., 42

K
Keller, A. E., 17
Kelso, J. M. R., 42

L
Lam, D. C. L., 42

M
Malanchuk, J. L., 4
Moreau, G., 28
Morgan, E. L., 84
Mundy, P. A., 4

N
Nesse, R. J., 4

P
Parent, L., 28
Perry, T. E., 67
Planas, D., 28
Pollman, C. D., 67

R
Ross, J. W., 92

S
Synnott, J. C., 92

W
Weaver, T. P., 84

Y
Young, R., 92

Subject Index

A

ACID (Acidification Chemistry Information Data Base), 7, 8
 Environmental Protection Agency Storage and Retrieval (STORES), 8
Acid deposition (*See* Acid rain)
Acid neutralizing capacity (ANC), 67, 72, 74(table), 76(illus), 77(table)
 effects of ion exchange
 biological sulfate reduction, 80, 81(illus)
Acid rain, 4, 17, 84
 aluminum salts
 in laboratory toxicity tests, 92, 96
 effects on aquatic biological systems, 1, 2, 4, 29, 39, 98
 monitoring, automated, 85, 88
 National Acid Precipitation Assessment Program (NAPAP), 88
 soft-water lake sediments, 67
 trends, study by National Academy of Sciences, 8
Acidification, 18, 19, 29, 34, 35, 42
 automated biomonitoring, 85
 effect on benthic decomposition rates, 19
 experimental, 69
 on biomass, 28, 38
 analytical methods used, 31(table)
 continuous impact, 37(illus), 38, 39
 short-term impact, 36(illus), 38, 39
 streams, 2, 29

watershed characteristics in relation to, 10(illus)
Algae (*See also* Ecosystems), 17, 42, 44
 acidification, 35
 benthos grazing, 38
 controlled by phosphorous concentrations, 18
 periphytic, productivity of running water, 28, 35, 38
Alkalinity, 2, 49
 in soft water lakes, 68, 69, 82
 effect of sulfate-reducing bacteria, 80
Aluminum
 effects on aquatic life, 69, 93, 94
 experimental concentrations, 30, 39
 influence of temperature on toxicity, 95
 salts in laboratory toxicity tests, 2, 92, 96
ANC (*See* Acid neutralizing capacity)
Angiosperms, 22
Anion effects (*See* Ion exchange)
Aquatic biological communities (*See also* Ecosystems), 1, 2, 23, 24
 effects of acid deposition, 4, 39, 98
 comparison of acid alone to acid plus aluminum, 29
 environment, 94
 food chain, 44
Autotrophic communities (*See also* Ecosystems), 2, 17
 Benthic, 19, 21, 38
 littoral, effects on ccosystems, 17, 22, 29

Autotrophic communities (*See also* Ecosystems) (*continued*)
  sediments, 74(table, illus)
  macrophytes, 19, 22, 23
  oligotrophic waters, 38, 47
  pelagic linkages, 17
    sediments, 74(table, illus)
  vegetation, influence on fish communities, 23, 24

B

Bacteria, 17
  sulfate reducing, 80
Benthic autotrophic communities (*See* Autotrophic communities)
Biological sensing systems
  automated monitoring of water resources
    system design, 85, 89, 90
    field trials, 87
Biomass (*See also* Ecosystems)
  effects on fish communities, 18
  experimental acidification on, 28, 35, 39
  possible explanation for increase, 38
Biomonitoring, automated, 2, 84
  system design, equipment and methodology, 85, 89
  field trials, 87
Biosensing (*See* Biomonitoring)
Biota (*See* Aquatic biological communities, Ecosystems)
Blue-greens (*See* Periphyton)
Breath detectors, fish, 85, 89, 90
Bryophita, 22
Buffering, 67, 79(illus), 80, 82
  metal ions in aquatic environment, 94

C

Carbon uptake, 42, 44, 45(illus), 49
  of phytoplankton, 49, 51
Chlorophyll-a measurement, 38(illus)

Copper
  in experimental acidification, 34

D

Data collection platform (DCP) sites, 87–89
Data retrieval, automated
  receiving and processing, 90
  system design
    field trials, 87
    remote data collection platforms (DCP)
    transmission to, from NOAA Geostationary Operational Environmental Satellite (GOES), 86
Diatoms (*See* Periphyton)

E

Ecosystems, 20, 21(illus), 30, 39
  effects of acidification in, 4, 68, 74
  vegetation, 23, 24
Electrodes, 98
Emissions, 4
Environmental factors
  biological systems as integrators of, 85
Eutrification, 2

F

Fish
  aluminum toxicity to, 69, 95
  bioassay, 92
  breathing responses detector, 85, 89, 90
  effects of acid deposition, 1, 2, 5, 9, 11, 13, 14(illus)
  electrolytes, 94
  measurement of physiological and behavioral responses, 90
Fish Information Network, 9
Fish resources, 17, 23, 24

SUBJECT INDEX

economic value of reduced resources, 14, 15
Fulvic acid, 94

G

Groundwater flow, 43, 50, 51(illus)

H

Hydrogen electrode, 98
Hydrogen ions
  difficulty of separating from metals, 30
Hydrology, 2, 42, 46
  model, 42, 43, 47, 52
  water chemistry gradients related to, 49

I

Invertebrates, stream
  growth rates
    in acidified streams, 59(illus), 60(illus)
    in laboratory conditions, 59(illus), 62(illus), 63(illus)
    temperature related, 60(illus), 61(illus), 63
  mortality
    in laboratory conditions, 56, 57
    in relation to size, 59, 62
  nutrients, 56
  sensitivity
    to acidification, 65
    to PH depression, 54, 57, 58(illus)
Ion exchange, 2
  anion effects, associated with use of aluminum salts in laboratory toxicity tests, 92
  effects of ANC in soft water lakes, 68, 69, 72, 75(table), 76(illus), 78, 82

K

KCl (*See* pH measurement)

L

Lake/stream acidification (*See also* Autotrophic communities, Watershed communities, Invertebrates, stream), 1, 10(illus), 11, 30, 56
  alkalinity levels, 49
  biological productivities, 47
  carbon dioxide, 49
  carbon uptake rate, 43, 51
  effects on fish populations, 17
  groundwater flow, 43
  limnological indicators, 47
  littoral-pelagic linkages, 17, 19
  monitoring, automated, 85
  NAPAP research program, 4, 7
  nutrient concentrations, 46(illus)
  under laboratory conditions, 56
  rate of acidification, 11
  sediments
    soft-water lakes in Florida, 67
Limnology, (*See also* Biomass, Ecosystems, Lake/stream acidification, Water quality), 17, 19, 47
Littoral autotrophic communities (*See* Autotrophic communities)

M

Macrophytes (*See also* Autotrophic communities, Ecosystems), 17, 19
Metal ions
  chemistry modified by fulvic acids, 94
Metal salts
  toxicity inversely related to solubility, 92

Meteorological conditions (*see* Watershed communities)
Mineral dissolution, 68, 69, 73
Models
    acidification to eutrification, 52

N

NAPAP (U.S. National Acid Precipitation Assessment Program)
    description of research program, 4–5
    organization chart, 6
National Surface Water Survey (NSWS), 7
    water quality parameters, 8(table)
Nitrite concentrations, 46(illus)
Nitrogen oxide emissions (*See* Emissions)
NOAA Geostationary Operational Environmental Satellite (GOES), 86
Nutrient concentrations
    in lakes/streams, 46

O

Ocean and sea linkages (*See* Autotrophic communities)
Oligotrophic waters (*See* Autotrophic communities)
Osmoregulation, 94
    in acid and acid/aluminum stressed fish, 95

P

Pelagic linkages (*See* Autotrophic communities)
Periphyton, 28, 35, 38
pH conditions, 18, 19, 23, 29, 39
    affected by sulfate load, 45, 47
    alkalinity, 47, 48(illus), 82
    buffering capacity, 67, 74(table, illus), 95
    depression, sensitivity of stream invertebrates, 54, 57, 58(illus)
    effects on aquatic life, 93
    influence of temperature on toxicity and metabolism, 95
pH measurement
    electrodes with low-resistance sensing elements, 99
    accuracy, 100
    description of method development, 100
    summary of method, 105–106
    Ross model, 101, 102(table), 103(table)
    provides fast and accurate measurement
    standard hydrogen model, 100(table), 105(table)
    response to temperature excursion, 101(illus)
    in acid rain, 3, 11
    method, 98
    in laboratory streams, 55
    in soft water lakes, 70, 71, 75(table), 77
    reproducibility in calibration buffers, 103(table), 104(table)
    improved by use of KCl to increase conductivity, 105
    temperature dependence of calibration buffers, 104(table)
Phosphorous concentrations, 18, 22
Photosynthesis, 49
    stimulation after short-term acidification, 38
Phytoplankton (*See also* Biomass, Ecosystems), 2, 17, 29, 47
    carbon uptake, 49, 51
    effects of acidification, 42, 47
Plankton (*See also* Ecosystems), 18
    limited by lack of nutrients, 2, 19

Precipitation runoff, 50, 51(illus)
Predation assessment
  by measurement of chlorphyll-a, 37

R

Real-time biosensing, 85
Remote biosensing
  computer satellite data retrieval, 84, 89, 90
  water quality monitoring, 87
Ross pH electrode, 3, 98, 101–103
Running water (*See* Acidification, experimental)

S

Satellite data retrieval (*See* Biomonitoring)
Sediments, 22
  laboratory experiments and methodology, 72, 73, 76
  in soft water lakes, 67, 68, 70(illus), 71(illus), 77(table)
  littoral-pelagic, 82
  reductions, 70
Silt (*See* Sediments)
Sphagnum, 22
Stream invertebrates (*See* Invertebrates, stream)
Streams, acidification (*See* Lake/stream acidification)
Stress responses of fish (*See also* Biomonitoring)
  osmoregulatory process, 94
  pH/aluminum, physiological effects, 93
Sulfate concentrations, 8
  reduction, 67, 68, 69, 78(illus), 80, 81(illus), 82
Sulfur emissions, 4
Sulphuric acid
  experimental acidification, 31
Surface water chemistry, 9, 10

T

Toxicity tests, 95
  salt free aluminum hydroxide precipitate for, 96
  use of aluminum salts, potential problems
    associated anion effects, 2, 92
    effects on fish, 69
Trophic-level interactions, 17
Turkey Lakes watershed (*See* Watershed communities)

U

Ultraoligotrophic waters (*See* Aquatic biological communities)
U.S. National Acid Precipitation Assessment Program (*See* NAPAP)

V

Vegetation (*See* Algae, Ecosystems)

W

Water chemistry, 2, 4, 7, 10(illus), 11
  algal growth tied to, 42
  littoral zone influence on, 22
  linked to fish presence/absence, 11, 12, 13(illus)
Water quality, 7, 9, 19, 20
  remote monitoring, automated, 84, 85, 87(illus), 89
Watershed communities, studies
  acidification, 52
  Adirondak Mountains, 7, 9
    effects of acidification, 11
    fish resources related to acidification, 13, 14(illus)

Watershed communities, studies (*continued*)
  Blue Ridge Province, 7
  Florida, 67
  New England, 7
  New York, 7
  Pennsylvania
  sampling stations, 44(illus)
  Turkey Lakes, Canada, 42, 43, 48, 51
  Upper Midwest, 7

Watershed hydrology (*See* Hydrology)
Wildlife management, 11

Z

Zinc
  in experimental acidification
Zooplankton (*See also* Aquatic biological communities, Ecosystems), 17, 22, 23